CLIMATE SECURITY

CLIMATE SECURITY

Ashok Swain

1 Oliver's Yard
55 City Road
London EC1Y 1SP

2455 Teller Road
Thousand Oaks
California 91320

Unit No 323-333, Third Floor, F-Block
International Trade Tower
Nehru Place, New Delhi 110 019

8 Marina View Suite 43-053
Asia Square Tower 1
Singapore 018960

Editor: Michael Ainsley
Editorial assistant: Pippa Wills
Production editor: Sarah Cooke
Copyeditor: Ritika Sharma
Proofreader: Girish Sharma
Indexer: TNQ Tech Pvt. Ltd.
Marketing manager: Fauzia Eastwood
Cover design: Francis Kenney
Typeset by: TNQ Tech Pvt. Ltd.

© Ashok Swain 2025

Apart from any fair dealing for the purposes of research, private study, or criticism or review, as permitted under the Copyright, Designs and Patents Act, 1988, this publication may not be reproduced, stored or transmitted in any form, or by any means, without the prior permission in writing of the publisher, or in the case of reprographic reproduction, in accordance with the terms of licences issued by the Copyright Licencing Agency. Enquiries concerning reproduction outside those terms should be sent to the publisher.

Library of Congress Control Number: 2024938688

British Library Cataloguing in Publication data

A catalogue record for this book is available from the British Library

ISBN 978-1-5296-7085-1
ISBN 978-1-5296-7084-4 (pbk)

CONTENTS

About the Author vii

Preface and Acknowledgement ix

1 **Introduction: The Climate Crisis and International Security**   1

2 **Climate Apathy and Self-Preservation**   23

3 **Global Militaries and the Climate Crisis**   47

4 **Shifting Territories and Borders**   69

5 **Water Conflict and Cooperation**   89

6 **Climate Migration**   107

7 **Regime Legitimacy and Democracy**   125

8 **Conclusion**   143

References   155

Index   179

ABOUT THE AUTHOR

Ashok Swain is a Professor and Head of Department of the Department of Peace and Conflict Research. He is the UNESCO Chair on International Water Cooperation and the Director of Research School of International Water Cooperation at Uppsala University. He is also the founding Editor-in-Chief of 'Environment and Security' journal, jointly published by Sage Publishing and Environmental Peacebuilding Association.

PREFACE AND ACKNOWLEDGEMENT

Climate change stands as one of the most formidable challenges of our time, intertwining with global security in complex and unprecedented ways. This book, 'Climate Security', delves into the multifaceted relationship between our changing climate and the security of nations and individuals alike. It explores how the increasing intensity of global environmental stress, spurred by human activities, is reshaping the landscape of global security, affecting countries, societies and individuals across the globe, particularly in the Global South.

The phenomenon of climate change is not a distant future threat; it is a present reality. Human-induced alterations in the Earth's climate system are manifesting through a massive build-up of greenhouse gases, leading to rising temperatures, sea-level rise and extreme weather events. These changes are not uniform; they impact different regions in varied ways, influenced by geographical, socio-economic and political factors. This book meticulously examines how different countries and societies are grappling with these changes, highlighting the role of cultural norms, social practices and institutional capabilities in shaping their responses.

The dire consequences of climate change are disproportionately borne by the poorer and developing economies, where reliance on natural resources for livelihoods is high and adaptive capacities are low. The narrative expands on how climate change exacerbates poverty, triggers migration and heightens the risk of conflict, especially in regions plagued by poor governance and fragile institutions. Through detailed analysis, this book illuminates the intricate ways in which climate change acts as a catalyst for conflict over resources, disrupts economic growth, erodes social cohesion and challenges the legitimacy of governments. In the realm of global security, climate change is prompting a re-evaluation of traditional military paradigms, pushing the boundaries of national security to include environmental considerations. The narrative explores the strategic implications of climate-induced changes in shared spaces like the oceans and polar regions. The emerging security challenges in these domains call for innovative approaches to diplomacy, defence and security strategies.

Many researchers and advocacy groups avoid framing climate change as a national security concern to prevent governments from prioritising military and border control spending over environmental action. This approach, while well-intentioned, has led to climate change often being sidelined in political agendas and policymaking. This book argues that climate change poses significant threats to national security, such as impairing military operations, altering borders, exacerbating natural disasters and potentially fuelling conflicts over resources and refugee movements. By highlighting these threats, the book aims to elevate the importance of climate change on national agendas, ensuring that political leaders take decisive action to mitigate its impacts. The ultimate goal is to balance the focus on both human and national security to address climate change effectively.

As we navigate this uncharted terrain, the book underscores the importance of international cooperation and collective action in addressing climate change. It emphasises that while individual countries can make significant strides, the global nature of the challenge necessitates a concerted, collaborative effort. Through a blend of rigorous analysis and forward-looking insights, 'Climate Security' aims to foster a deeper understanding of the interplay between climate change and security, providing a roadmap for countries to prepare and strengthen their resilience against the impending challenges. In essence, this book is a clarion call to acknowledge and act on the inseparable link between climate change and security. It seeks to galvanise policymakers, scholars and the global community to view the climate crisis not just as an environmental or economic issue but as a pivotal security concern that demands immediate and sustained attention. As we stand at the crossroads, the choices we make today will determine the security and well-being of generations to come.

In crafting 'Climate Security', I have been privileged to garner the support, wisdom and insights from a diverse group of individuals and institutions, each leaving a distinct and enduring imprint on this book.

I would like to thank Maria Båld for her capable research assistance, hard work, and excellent support in reviewing literature for this book project. I am deeply grateful to my colleagues at the Department of Peace and Conflict Research, particularly those at the Research School of International Water Cooperation, including Florian Krampe, Kyungmee Kim, Charlotte Grech-Madin, Stefan Döring and Jonathan Hall, for their valuable insights and discussions that have expanded this book's horizons. Their analysis of the nuanced interplay between the environment and security has been indispensable.

I would like to extend my heartfelt thanks to longstanding co-authors and friends, such as Joakim Öjendal, Anders Jägerskog, Roland Kostic, Ken Conca, Geoff Dabelko, Larry Swatuk, Stacy VanDeveer, Juliano Cortinhas, James R. Lee, Giuliano Di Baldassarre, Aaron Wolf, Lauren Herzer Rishi, Prakash Kashwan, Dan Smith, T.V. Paul, Anders Themnér, Lara B. Fowler, Shafiqul Islam, Saleem Ali, Anamika Barua, Adeel Zafar, Fredrik Söderbaum, Sofie Hellberg, Michael Schulz, Jeremy J. Schmidt, Cameron

Harrington, David Mwaniki, Cedric De Coning, Thor Olav Iversen, Amit Prakash, Sajid Karim and Adil Najam, with whom I've had the pleasure of collaborating on research over the years.

Additionally, I would like to express my gratitude to my editor colleagues of the journal 'Environment and Security' – Carl Bruch, Tobias Ide, Päivi Lujala, Richard A Matthew, Erika Weinthal and Tom Delgiannis – for their intensive and enriching discussions on environment, climate change and security topics in recent years.

I also would extend my appreciation to environmental organisations and security think tanks such as SIWI, SIPRI, the Woodrow Wilson Center and NUPI for granting access to essential data and reports, thereby significantly enhancing this manuscript's quality and authenticity.

Special thanks are due to my editor, Michael Ainsley, who not only persuaded me to undertake this project but also provided meticulous attention to detail and continuous support throughout the writing process, serving as a beacon of guidance. I am also thankful to the editorial assistant, Pippa Wills, and the publisher, Natalie Aguilera, of Sage for overseeing the production process of the book.

I express gratitude to my family for their support throughout the journey of writing this book.

Lastly, my profound gratitude goes to my academic mentor, Professor Peter Wallensteen, whose encouragement to delve into water and environmental issues post-doctorally has been a cornerstone of my academic journey and research for over three decades. This book is dedicated to him; a testament to his enduring influence and inspiration.

This book embodies the collective efforts of many who are ardently committed to fostering a sustainable and secure future. To everyone who has played a role, whether directly or indirectly, in this endeavour, I extend my deepest appreciation.

Ashok Swain, Uppsala

1
INTRODUCTION: THE CLIMATE CRISIS AND INTERNATIONAL SECURITY

Climate change isn't just knocking at humanity's door; it's barging in with full force, presenting one of the gravest challenges we've ever faced. What once seemed like a distant environmental concern has rapidly morphed into an urgent security threat, one that shakes the very foundations of global stability. This isn't just about polar bears or melting ice caps anymore. It's about geopolitical upheaval, dwindling resources and the well-being of people everywhere. We're staring down the barrel of an environmental catastrophe, one that amplifies every threat and challenges our societies grapple with.

As we navigate this treacherous landscape, it's clear that going alone isn't an option. Only by working together can we hope to tackle the mammoth task ahead. Political leaders need to wake up to this reality and put climate action at the top of their agendas. If they don't, we risk losing not just our environment, but our very ability to survive as societies. To get a grip on this crisis, this book begins with understanding the concept of security – how it's evolved and how it's intimately tied to the climate. In the first chapter, we dive into this exploration, shedding light on the shortcomings of our current approaches to climate change and arguing passionately why climate security must be seen as a matter of national urgency.

WHAT IS 'SECURITY'?

The concept of security has evolved significantly over time, reflecting changes in political, social, economic and technological landscapes. Historically, security primarily referred to the protection of a state's sovereignty and territorial integrity from external military threats. This concept gained particular prominence during the Cold War era. During that time, the world witnessed an intense focus on military capabilities, nuclear deterrence and state-centric security concerns. This period was marked by a bipolar power structure, with the United States and the Soviet Union at its core. Security studies and policies were predominantly centred on protecting national borders and deterring military aggression. The primary goal was to maintain state sovereignty and political independence against potential external threats.

The dissolution of the Soviet Union and the end of the Cold War in Europe at the beginning of the 1990s marked a significant turning point in the understanding of security. The binary structure of global power dynamics dissipated, giving way to a more complex and interconnected world. This shift prompted a re-evaluation of traditional security paradigms. The focus started to move away from state-centric military threats to include a broader range of challenges. In the post-Cold War era, the world began to face diverse challenges that extended beyond the traditional military scope. The argument is that in the post-Cold War world, threats such as civil wars, transnational crimes, terrorism, cyber wars, hybrid wars and infectious diseases are more prevalent than interstate wars. Globalisation further complicates the security landscape. The interconnectedness of economies, cultures and politics across borders means that issues in one part of the world can have ripple effects globally. Globalisation also brought to the forefront the challenges of economic security, highlighting how economic crises could undermine national stability. In other words, the increasingly interconnected world has created an environment that fosters new challenges to countries striving for external and internal security.

While wars have become relatively rarer in the last 50 years, civil wars have become more common and intractable. Genocide, crimes against humanity, war crimes and various human rights violations continue to endanger peace and security, locally, nationally, regionally and globally. Racism and xenophobia are being used to divide people and societies, and the rising political and social polarisation leads to violence in countries or regions. Moreover, unconventional threats like climate change, environmental scarcity, large-scale human migration, social injustice, the spread of violent and hateful ideologies, food and water scarcity and increasing pandemics have posed serious challenges for the countries to achieve both external and internal peace. In some of my books, I have elaborated on how these non-military threats are also interrelated, and a threat to one country or region has often become a threat to all (Swain, 2013; Swain and Jägerskog, 2016). The realisation that these non-military threats could have devastating effects on national and global stability led to an expanded understanding of security. This broader perspective encompassed not just the safety of state territories but also the well-being of societies and individuals.

Though traditional security concerns like border protection, sovereignty and political stability remain crucial, with the end of the Cold War and the rise of globalisation, the traditional approach to security has been criticised for being too narrow. The Cold War era focused on a state-centric view, emphasising military strength. In contrast, more recent perspectives advocate a comprehensive approach, considering economic, environmental and social aspects, and recognising the interconnected nature of modern global challenges. Scholars differ in their views on security. The debate between 'traditionalists' and 'wideners' reflects this evolution in understanding security. Traditionalists focus on military aspects and state sovereignty, while wideners advocate for a more inclusive approach that recognises a variety of threats, including non-military ones, and considers the security of individuals and the global community (Swain, 2013). Security researchers like Barry Buzan and Richard Ullman challenged the traditional definition of security in the 1980s, advocating for a broader concept. Buzan questioned the appropriate level for assessing security, whether individual, national or international, while Ullman warned against a military-only definition, which could overlook other significant threats (Buzan, 1983; Ullman, 1983). This broader approach includes non-military issues such as ecological, economic, political and sociocultural factors, considering not just the state but also individuals and the international system. The concept of security has always been contested and elusive, with various interpretations influenced by moral, ideological and normative elements. The push for a broader understanding of security includes non-military issues and encompasses societal and individual levels. Overall, the concept of security is continuously evolving, reflecting the changing dynamics of global politics and the complex nature of modern threats. This ongoing debate underscores the need for a more inclusive and comprehensive approach to security in today's interconnected world. In this debate, the concept of human security emerged as a significant shift in the security discourse. This approach places the individual at the centre of security concerns, rather than the state. It emphasises protecting people from critical and pervasive threats to human rights, safety and even survival.

The United Nations Development Programme (UNDP) brought emphasis to the human security concept in 1994 (UNDP, 1994). The introduction of this concept paved the way for an approach shift from security through armament to security through sustainable human development and one that goes beyond the traditional emphasis on the territory. With the birth of the human security concept in the post-Cold War period, all cornerstones of the security assumptions centring on nation-states have been questioned. The concept of human security focuses on the safety of the individual and human economic, health, food, social and environmental needs. The initial argument was that people needed to have 'freedom from fear' and 'freedom from want', which is critical to national, regional and global security and peace (Hanlon, 2016). 'Freedom from fear' involves the protection of human beings, which includes threats that are directed towards individual safety, like armed conflict, terrorism, ethnic expulsion, illegal arms trade and political and criminal violence. These forms of threats are directly violent in nature. However, 'freedom from want' broadens the human security agenda further to include the freedom of not having to worry about basic needs such as experiencing poverty, hunger, diseases, natural disasters and displacement. It emphasises linking sustainable development with peace and security. This definition of human security has been further expanded by the Commission on Human Security by adding the freedom of future generations to inherit a healthy natural environment (Ogata and Sen, 2003). Human security thus applies to all sources of not only present but also future insecurity towards the individual, advocating for sustainable development.

In the light of human security, the world has committed to 17 Sustainable Development Goals (SDGs) since 2015 as a holistic approach to achieving sustainable development and human security. The SDGs range from eradicating all forms of poverty and calling for zero hunger, to ensuring peace, justice and strong institutions by 2030. Nevertheless, the number of people affected by hunger has increased since the COVID-19 pandemic (FAO, 2021a). The UN Department of Economic and Social Affairs, in its World Social Report 2020, points out that inequality is growing for more than 70% of the world's population, and that is increasing risks of societal divisions and adversely affecting socio-economic development (UN-DESA, 2020). Individuals, unlike a country, have a long and infinite list of core values, as the definition of human security emphasises: Besides political and socio-economic factors, the other key areas are environmental, food, health and group rights. Globalisation is generating new wealth and encouraging technological innovations, but at the same time, it has not been successful to support and promote sustainable human development. Many regions and societies have been left behind. They face real challenges for basic survival, which creates huge uncertainties about future peace and security in various parts of the world.

While the world is struggling to achieve human security worldwide, other types of security threats are also increasing. Despite the Ukraine War since 2022, interstate wars have reduced considerably in recent decades. Wherever countries are fighting against each other, conflicts rarely are fought between two rival armies on a battlefield. Taking a hybrid form and combining both military and non-military methods, the emphasis of warring countries is on civilian spaces in the new age wars. 'Hybrid threats' refer to multifaceted threats, ranging from hacking to terrorism, from disinformation to migration occurring in civilian spaces (Bilal, 2021). The new wars in their hybrid forms have become diffused, combining state and non-state actors. It has become almost impossible to differentiate the bilateral interactions, whether peaceful or confrontational. The key attraction for the states to engage in a hybrid war is hostile actions are largely non-attributable as they fight via proxy actors.

Cyberattacks between two adversaries have also increasingly become a common method of warfare. As human life is increasingly integrated into networks, and the 'Internet of Things' (IoT) expands, the threats through cyberspace are becoming more potent (Weimann, 2004). Countries are yet to witness cyber-terror from militant groups, but it can happen anytime. Terror groups have not yet targeted network infrastructure, possibly because they are also dependent on cyberspace to recruit new members and even manage their terror operations. However, a technologically savvy lone wolf causing massive damage to human life and properties with a cleverly planned cyber-terror attack is just a matter of time. Cyber methods have already become key weapons in hybrid wars where the focus is more on the civil population than the military. Though the countries have been invariably extra careful to protect their military communications and infrastructure from cyberattacks, societal vulnerabilities have become easy targets in the era of hybrid warfare. Disinformation spread through cyberspace is being used to agitate, radicalise or terrorise a country or a specific population group.

Besides countries dealing with new forms of warfare, several of them are also fighting civil wars within their territories. The number of internal armed conflicts is rising, and fatalities are also increasing. Though there are around 55 armed conflicts ongoing worldwide, at least eight of them are extremely violent, with more than 1,000 battle-related deaths per year (UCDP, 2023a). These conflicts are not limited to one particular region, but one common factor is that they are not new conflicts. If a civil war is sparked by ideology, there is less possibility of it being intractable as the change of regime can bring the termination of the conflict. However, the real challenge remains to find a long-term sustainable, peaceful resolution to a civil war driven by identity issues, such as religion, race, ethnicity or language. Unfortunately, nearly two-thirds of the world's civil wars are identity-based. Though the world is witnessing the birth of very few new conflicts, the real challenge to contain organised violence has been these identity-based armed conflicts, which are hard to find a negotiated solution unless it results in secession from the country. The civil wars are also witnessing more and more direct involvement of external forces. They provide both parties with material, moral and military support and help the conflict last longer.

The complexity in defining security arises from the evolving nature of threats, technological advancements, the interconnectedness of global systems and the need to balance diverse and sometimes conflicting interests. This complexity requires a multi-faceted and adaptable approach to security policy and strategy. Thus, defining security is a difficult proposition as it could be a process or outcome. The varied sets of norms and assumptions within these different components of security prevent it from being put under one definitional fiat (Buzan, 1984). As the concept of security keeps evolving with time, it is prudent not to confine it within a rigid rational structure (Heathershaw, 2008). A global authority on peace research, Professor Peter Wallensteen argues that the definition of security should not only narrow down to the absence of military or territorial threats to a state but also the presence of factors that make society more peaceful and secure (Wallensteen, 2011). The security of a country and society must be rooted in social justice, the ability of all to exercise their rights regardless of their gender, race, religion, ethnicity or nationality.

The new security threats to the world, irrespective of big or small, rich or poor, East or West and North or South, are mutually vulnerable. The increasingly ambiguous and inclusive nature of measuring security is now unavoidable. The world and its problems have become excessively complex and diffusive, demanding a more exhaustive approach to deal with such challenges. Finding an inclusive approach to foster comprehensive security has become a necessity just as maintaining security through nuclear or military deterrence has lost a major part of its significance. Armed conflicts and the threat of armed conflicts undermine security but so do numerous other environmental and socio-economic challenges that create tensions, drive inequalities and instability create the conditions. The newly emerging threats to security are not necessarily or conventionally armed in nature but have a global reach with severe and disruptive consequences for people's physical and psychological well-being and their ability to exercise their human rights.

Undoubtedly, the understanding of security and connected risks has become complex and comprehensive. Increasing military strength and spending is neither sufficient to win the new types of war nor capable of successfully meeting unconventional emerging security threats. Thus, the challenge for the countries is to forge a new inclusive and reflexive security architecture that is both effective and responsive in dealing with the complex challenges to global security in the 21st century. In the last 50 years, the world has become a 'global village', which has led to the creation of many non-state actors and forces that impact beyond the national boundaries of state actors. The new reality of achieving security presents several challenges. Perhaps, the most pressing one is the newly emerging threats and ways of fighting the wars as they are not conventionally armed in nature but have global implications. In short, security is being confronted with a new set of complex and interrelated risks. Thus, to achieve comprehensive security, the world must accept the importance of multilateralism and commit to intelligent management of social, political, economic and environmental resources, keeping the focus on human security.

Countries need to adopt a more complex and comprehensive approach to analyse their strategy to achieve peace and security beyond the traditional prism of military power. Increasing military strength and spending is neither sufficient to win the new types of war nor capable of successfully meeting unconventional emerging security threats. Thus, the challenge for the countries is to forge new sustainable and comprehensive security architecture. At present, the search for security should be a search for comprehensive security, in which global, national and human security and development are intertwined. Security is something that all human beings are entitled to, despite their race, ethnicity, gender or nationality, and can only be gained through mutual efforts and multilateral mechanisms. The inadequacy of the traditional concept of security begs the need for an understanding of security amenable to the 21st century. The approach that moves beyond the nation-state towards a broader and deeper analysis includes traditional challenges to security and economic, societal, cultural and environmental issues. An integrated approach to security can only be better equipped to provide sustainable peace and security to this changing world and its emerging challenges. However, in this inclusive framework, the state continues to remain the prime provider and negotiator of the security of its population and the planet.

THE CLIMATE CRISIS

One of the most prominent emerging non-military security challenges is climate change. Climate change can create new security challenges, while also exaggerating already existing ones. The human-induced climate crisis is heating our planet at an unprecedented pace. Although the climate has changed naturally throughout history, due to external and internal factors such as variations in solar activity and volcanic eruptions, it is unequivocal that human activities are responsible for the current rapid rise in global temperatures (Bogren et al., 2008; IPCC, 2023a). Extensive scientific research indicates that the current rate and extent of warming far exceed the rate that could be explained by natural variations. Instead, human activities, especially connected with the release of greenhouse gas emissions and extensive land use change, are dominant drivers of the current global warming. To differentiate between natural and human-induced climate change, the latter is often described as anthropogenic climate change, meaning that it originates from human activities. The anthropogenic climate change poses severe security challenges. Those who claim that human-induced climate change is not responsible for the changes in weather patterns seen today often rely on a combination of misinformation, selective data interpretation, or misunderstanding of climate change.

To understand how climate change leads to warmer temperatures, it is crucial to grasp the basic science behind global warming. The Earth's climate depends on a delicate balance of energy exchange between the Sun and the Earth. Somewhat simplified, solar radiation from the Sun reaches the Earth's surface, with some absorbed and later re-emitted as heat. The atmosphere surrounding the Earth consists of various gases, such as carbon dioxide (CO_2) and methane (CH_4) which ensure that some heat from the sun can warm up the planet sufficiently for life to exist. Changes in the chemical composition of the gases, such as increasing amounts of CO_2 and CH_4 trap more heat in the atmosphere, causing a gradual warming of the planet (Bogren et al., 2008). Therefore, human activities consisting of fossil fuel production and land use changes emitting vast amounts of greenhouse gas emissions causing an increase in global temperatures. The rising global temperatures due to human activities clearly set off from the Industrial Revolution onwards. Now, the planet is warming up dangerously while greenhouse gas emissions are not showing any signs of slowing down (WMO, 2023a). Still, political leaders have not committed to any credible action plan that would ensure rapid emission reduction to limit global warming.

The latest consecutive nine years (2015-2023) have been the warmest years on record (WMO, 2023a). Of these, 2023 was the warmest year ever recorded since the observations started 174 years ago (WMO, 2023a). During the year, extended periods of heatwaves have caused severe implications for public health, agriculture and ecosystems. The high temperatures also contributed significantly to wildfires in many parts of the world, releasing heat-trapping gases and elevating local temperatures (Copernicus, 2023). In Canada alone, wildfires burned a record-breaking 18.4 million hectares in 2023, in contrast to the annual average of wildfires in Canada which is around 2.5 million hectares (NASA, 2023a). The devastating wildfires have serious societal impacts with loss of lives and destruction of homes.

As the Earth warms, it can lead to changes in atmospheric circulation patterns, making the weather less predictable and more extreme. In turn, this can cause prolonged periods of heat, drought, more intense storms and flooding events (IPCC, 2023a). These events can have serious impacts on societies, through reduced water and food security, destruction of homes and livelihoods, and can cause population displacement. The rising temperatures are also causing glaciers and ice caps to melt rapidly, contributing to sea-level rise. For instance, the glaciers in North America and the European Alps have experienced extreme melting seasons in recent years. Moreover, oceans worldwide are warming up rapidly. A large majority of the heat that is trapped on Earth is stored in the oceans. In turn, this leads to changes in the ocean currents, rising sea-levels and contributes to changes in weather patterns, among other consequences (IPCC, 2023a). For instance, warmer oceans transport more heat and moisture into the atmosphere, which can fuel intense and prolonged heat waves and other extreme weather events.

Eventually, the concentration of greenhouse gases in the atmosphere and ocean, changes in water patterns, land use changes and biodiversity loss risk reaching tipping points where irreversible change in the Earth's systems is inevitable, and the possibility for humans to live in a 'safe operating space' is reduced (Richardson et al., 2023). In particular, climate change can trigger feedback loops amplifying global warming, which can contribute to the risk of transgressing tipping points. For instance, thawing permafrost releases stored greenhouse gases, such as methane and CO_2 into the atmosphere. In turn, the emissions contribute to the intensified greenhouse effect, leading to warmer temperatures and increasing permafrost thawing. Eventually, tipping points may be reached which can bring irreversible change to the previously relatively stable environmental conditions, such as the permafrost. Already now, several regional climate tipping points of Earth systems that are important to stabilise the planet's climate have been transgressed which weakens the global resilience capacity (Mckay et al., 2022). In other words, global warming combined with other environmental issues has various effects on our planet. In turn, environmental changes can pose serious risks to societal security, justice and peace. The challenge is therefore not only the warming temperatures, but the disruption it has in ecological, social and economic systems. It can contribute to making livelihoods less secure, impede economic growth and development, undermine social cohesion and increase competition for natural resources. Already now, the world is experiencing an increasing number of devastating natural hazards, which are expected to increase in the future.

Moreover, the pattern of global warming is uneven around the globe, already affecting some regions more than others. For instance, the Arctic region is warming faster than the rest of the world. In the Arctic, the mean temperature has risen almost 3°C compared to pre-industrial levels, warming four times faster than the global average (Rantanen et al., 2022). The rapidly warming Arctic region has serious geopolitical implications as the vast amount of more easily accessible natural resources in the area can increase tension between countries that compete for the rights to exploit the resources. In the Middle East and North Africa region, the temperature is warming twice as fast as the global average. By 2050, the regional average could be 4°C warmer than pre-industrial levels (WEF, 2023a). Undoubtedly, the arid region risks facing severe vulnerabilities connected to extreme heatwaves when the regional average temperature increases quickly. This can lead to more frequent and intense water and food insecurities, environmental degradation and desertification. In turn, this can aggregate already existing inequalities, trigger large-scale migration and provoke tensions and disputes within and between countries. Extreme temperatures and severe droughts have already affected economic activities and agricultural production in the region, in some cases spurring violent protests (Lindvall, 2021). In other words, climate change is an increasingly critical risk that already poses severe security challenges for our societies.

'CLIMATE SECURITY' AS A CONCEPT

Climate change's unprecedented impact on global weather patterns has far-reaching security implications on various levels. Unlike traditional security threats that often can be contained or localised, climate change is insidious and far-reaching, affecting every corner of the globe. In this context, the concept of climate security has emerged due to the increasing recognition of the profound and far-reaching impacts of climate change on global stability and security. The origin of climate security can be traced back to the late 20th century when environmental issues began to gain prominence on the global stage. Throughout the 1980s and 1990s, the concept of environmental security began to evolve, linking environmental issues with national and international security. The end of the Cold War and the rise of globalisation contributed to a broader understanding of security, beyond traditional military concerns. This period witnessed the recognition of non-traditional threats, including environmental degradation, as factors that could destabilise nations and regions (Swain, 2013).

The early 21st century marked a pivotal moment in the evolution of climate security. The increasing evidence of climate change and its potential impacts on global stability brought the issue into the security discourse. Key events, such as the release of the Intergovernmental Panel on Climate Change (IPCC) reports, played a significant role in highlighting the urgency of the issue. These reports provided scientific evidence of climate change and its potential to exacerbate resource scarcity, natural disasters and displacement of populations. The concept of climate security gained institutional recognition with various international organisations and national governments integrating it into their policies. In 2007, the United Nations Security Council held its first debate on climate change as a security issue, acknowledging the potential of climate change to exacerbate existing threats to international peace and security (UN, 2007). The next open debate took place in 2011 and resulted in a presidential statement expressing concerns that the adverse effects of climate change may 'aggravate certain existing threats to international peace and security' (UN, 2011). This period also saw the integration of climate considerations into national defence and security strategies, recognising the need for adaptation and mitigation measures. Climate adaptation measures signify strategies to adjust to the effects of climate change, such as investing in measures to reduce the impact of flooding. Climate mitigation measures are strategies to reduce the severity of climate change by decreasing the amount of greenhouse gas emissions released into the atmosphere.

Still, the Security Council has yet to reach a consensus on incorporating climate-related security risks into its regular business. Among the permanent members, the United Kingdom and France have been supporting this initiative from the beginning. Under the Biden administration, the United States has become a prime supporter of pushing the Security Council to include climate change within its agenda (Ryan, 2023). Two veto-carrying members, China and Russia, have openly taken a stance against it. China has for a long time argued that climate change is an issue of sustainable development, not a security issue (Moore and Melton, 2019). Russia's opposition has not been as rigid but still sceptical about including climate change as it fears that it will stretch the mandate of the Security Council (Krampe and de Coning, 2021).

Despite their opposition to formally agreeing on bringing climate change to the Security Council agenda, Russia and China have taken various measures on their domestic fronts to prepare themselves against climate-induced security challenges. They have also agreed to include the implication of climate change in different Security Council resolutions on several UN field missions since 2017, including on Lake Chad, Darfur, Somalia, Central Africa, Mali, DRC and recently Iraq.

Non-state actors, including non-governmental organisations (NGOs), academia and the private sector, have played a crucial role in advancing the concept of climate security. These actors have been instrumental in conducting research, raising awareness and advocating for policy changes. Their efforts have contributed to a more nuanced understanding of the complex interlinkages between climate change and security. Climate security intersects with a range of other global issues, including human rights, development and migration. The impacts of climate change, such as extreme weather events and sea-level rise, have implications for human rights and can exacerbate socio-economic inequalities. At the same time, climate-induced migration is emerging as a significant concern, with populations forced to move due to changing environmental conditions.

Over the past two decades, research has increasingly highlighted the complex interplay between environmental stress, exacerbated by climate change, and societal insecurity, often leading to armed conflict. The scarcity of renewable resources and the forced migration due to resource depletion are identified as key drivers of potential violent conflicts. Climate change is leading to severe consequences like sea-level rise, which threatens densely populated low-lying areas and island nations, altering typical rainfall patterns and increasing the frequency of extreme weather events. These changes not only affect environmental conditions but also have profound implications for human security, particularly in developing countries where agricultural dependency is high and adaptive capacities are low. Climate change is often considered a 'threat multiplier' because of its potential to aggravate existing vulnerabilities and increase the potential for conflicts (Swain and Öjendal, 2018). Mainly, because changing climate is not always a threat in itself, but its consequences can accelerate already existing challenges, particularly in regions with weak governance and limited resources to tackle the increasing issues. The international discourse on climate change has been marked by disagreements and a lack of consensus on mitigation strategies, especially between nations with different economic capacities, hampering efforts to address this global crisis effectively. This ongoing debate underscores the challenges in reconciling different national interests and lifestyles, and the urgent need for cooperative global action to mitigate the impacts of climate change.

There is no doubt that climate change has become a serious concern for this planet's security and even survival. Climate change is having a tremendous impact on the livelihood of societies and poses challenges to the peace, security and stability of countries at various levels. In the early decades of its inception, the Security Council solely focused on war, armed conflict and military threats. With the gradual broadening and widening of the security as a concept, the ambit of the Security Council focus has also expanded. In recent decades, its resolutions have covered human security challenges, and the protection of women and children from various pandemics and natural disasters. In this context, bringing climate change to the Security Council's agenda can be seen as a natural development.

The evolution of climate security from a marginal environmental concern to a central component of security policy underscores the changing nature of global threats. The recognition of climate change as a security risk reflects a broader understanding of the interconnectedness of global challenges. As the world grapples with the realities of a changing climate, the concept of climate security will continue to shape policies and strategies aimed at ensuring a stable and sustainable future. The journey from recognising the environmental impacts of human activity to addressing the security implications of climate change marks a significant shift in how we understand and respond to global challenges. Moreover, the emergence of the climate security concept is a response to the growing understanding of the multifaceted impacts of climate change, which extend beyond environmental degradation to encompass broader socio-economic and geopolitical dimensions. It reflects a paradigm shift in how security is perceived and addressed in the face of global environmental challenges.

Looking ahead, the concept of climate security is likely to evolve further as the impacts of climate change become more pronounced. There is a growing recognition of the need for a comprehensive approach to address the multifaceted challenges posed by climate change. This includes enhancing resilience, promoting sustainable development and fostering international cooperation to mitigate the security risks associated with climate change. The path forward for climate security is fraught with challenges but also presents opportunities. Addressing the security implications of climate change requires coordinated action at multiple levels, from local to global. It also requires bridging the gap between various disciplines, including environmental science, security studies and public policy. However, this challenge also presents an opportunity to foster a more integrated and holistic approach to security, one that accounts for the complex interdependencies of the modern world.

CLIMATE SECURITY AS NATIONAL SECURITY

In the 1970s, the Club of Rome's Project on the Predicament of Mankind emphasised the need for a planned and careful transition from economic growth to sustainability and visualised that availability and unavailability could affect the growth of global natural resources (Meadows et al., 1972). Providing food and shelter to a rapidly expanding world population has led to the widespread devastation of renewable resources like land, air, water, forest and biodiversity. The decline in the availability of these resources already threatens the life and survival of present and future generations. Climate change has brought further uncertainties over the availability of these resources and even threatens the survival of this planet. There is no doubt that climate change has become a serious concern for this planet's security and even survival. Climate change is having a tremendous impact on the livelihood of societies and poses challenges to the security and stability of countries at various levels.

One of the most immediate threats posed by climate change is resource scarcity, particularly concerning water and food (Swain, 2015). Ensuring food security for a large number of countries has become enormous. In the last 50 years, the world population has doubled from four billion in the mid-1970s to almost eight billion in 2022. Increasing living standards and changing food habits have also added the need for more food production. However, the diversion of agricultural land for urban expansion, biofuel production and declining water availability pose serious risks to the food security of many countries. Thus, there is an urgent need to boost agricultural production in sustainable ways that do not negatively impact the economy or the environment. While previous food shortages were mitigated by increasing agricultural land acreage, a booming population with increased urbanisation and less water supply makes this option less feasible in several countries.

Water is critical for human survival, economic development and peace and security. Water scarcity in the world has grown manifold in the last 50 years. The world is at present experiencing a severe global water crisis. More than half of the world's population is suffering from water scarcity at least some parts of the year (IPCC, 2022a). Climate change and economic developments have also brought changes to water supply and demand patterns. Countries have started disputing more over their shared rivers. There are at least 310 international river basins and 592 international groundwater basins in the world. The existing arrangement of water resources between and within countries in the arid and semi-arid regions has become more conflictual.

As some of the countries become increasingly uninhabitable due to climate change, there is a potential for significant geopolitical upheaval. The sea-level rise and melting of glaciers are moving national borders. The other primary concern is climate-induced migration, where large populations are forced to move in search of habitable land and resources. Large-scale population migration poses a severe challenge to the peace and security of many nations globally. Armed conflicts are forcing people to move to places searching for safety and economic opportunities attracting many to migrate to seek better living standards. It is not new that trans-border migrants, in some cases, support the ongoing conflict or act as spoilers of peace negotiations in homeland conflicts.

In other instances, they positively contribute to conflict resolution processes in their homelands. However, what is new is migration forced by climate change is becoming one of the world's major challenges. For peace and security, the countries need to be prepared to face the growing challenges of climate-induced forced migration.

Climate change has increasingly become a national security issue due to its profound impact on stability and safety from local to global levels. It can exacerbate already existing problems like resource scarcity, economic instability and geopolitical tensions. Changes in climate can lead to more frequent and severe natural disasters, such as hurricanes, floods and droughts, straining a nation's emergency response capabilities and infrastructure. As ecosystems and weather patterns shift, countries may also face new challenges in maintaining the integrity of their borders and managing increased competition for scarce resources. The destabilising effects of climate change can therefore undermine security by increasing the likelihood of conflicts, disrupting global supply chains and requiring significant adaptation and mitigation efforts by governments. This situation demands a strategic response from security apparatuses to anticipate, prepare for and respond to the wide-ranging impacts of climate change. However, the discourse and debate on climate change must emphasise the serious threats climate change poses to national security to get the issue to the top of the country's policy agenda. Climate change represents one of the most complex and multidimensional security threats of the 21st century. Its impacts are far-reaching, affecting geopolitical stability, resource availability, economic prosperity and social cohesion.

Highlighting how climate change is seriously affecting a country's military's strength and preparation is a low-hanging fruit for the political leadership to elevate climate to a national security issue. The hotter temperature has made difficult conditions for military missions and has adversely affected the viability and durability of military hardware. The changing climate has also exposed troops to new diseases and other health challenges. Increasing natural disasters like floods and hurricanes force authorities to divert troops from their regular border security duties to rescue and relief work. Rising sea-levels and extreme weather can damage infrastructure, including naval bases and other military installations. Moreover, militaries are increasingly called upon for humanitarian assistance in the wake of climate-induced disasters, stretching their capabilities and resources. The changing nature of the battlefield due to climate factors requires a re-evaluation of military strategies and preparedness.

Additionally, sea-level rise and increasing glacier melting have started to redraw the borders and disputes over exclusive economic zones. Putting the focus on climate change potentially changing the country's border is another surest way to demand the attention of the national security apparatus. Climate change also can create wars between countries over shared rivers and climate migrant movements.

Moreover, climate change, by bringing volatility to food prices and reversing economic growth, can also undermine social cohesion and government legitimacy leading to destabilisation of the political system, rebellion and military coups. There are many ways climate change poses serious threats to national security, and those need to be brought into the public and political discourse. Emphasis on climate change's impacts on national security doesn't necessarily mean ignoring its threats to human security but gaining greater attention from the political class to take urgent and concrete measures to limit global warming and prioritise adaptation measures in many climate-vulnerable countries. That will need the countries to consider climate issues as a national security challenge and to address these issues they need to be actively engaged in not doing it in an isolationist manner but cooperating with other countries and societies.

INADEQUACY OF MULTILATERAL AGREEMENTS

The climate crisis has been known for political leaders for a long time. Various attempts have been made to address the climate crisis in multilateral agreements. In the 1972 UN Conference on the Human Environment in Stockholm, the environment was for the first time discussed as a major issue in an international conference. Under the slogan 'Only one Earth', the states agreed to take steps to preserve and improve the human environment, by preventing pollution and reducing greenhouse gas emissions, among other things (UN, 1973). The 1972 Stockholm Conference became iconic for the modern environmental movement and aroused widespread interest and civic participation worldwide. Not only the global leaders but activists from all over the world also had gathered in Stockholm in a never-before-seen global movement for the environment. The final report from the conference included a clear warning:

> A point has been reached in history when we must shape our actions throughout the world with a more prudent care for their environmental consequences. Through ignorance or indifference, we can do massive and irreversible harm to the earthly environment on which our life and well-being depend. (UN, 1973:3)

But what has happened during these five decades? Unfortunately, not so much that world leaders or environmental activists can be proud of. Instead of heeding this warning, the world continues to overexploit nature, destroy the soil, pollute the water, cut down the forests and do very little to control the greenhouse gas emissions filling the atmosphere and ocean. Countries continue to ignore that environmental degradation and climate change take away basic human security.

Following the 1972 UN Conference on the Human Environment, several international conferences and negotiations have taken place, and information-sharing bodies have been created to strengthen countries' commitment to reducing emissions and ensuring healthy environments. In 1988, the UN founded the IPCC to provide scientific information about climate change that can be used as a basis to develop climate policies (IPCC, 2023b). A few years later, at the 1992 UN Conference on Environment and Development in Rio de Janeiro, the UN Framework Convention on Climate Change (UNFCCC) was adopted. In the same year, it was signed by 158 member states. The UNFCCC set a framework for international agreements to protect the global environment. The framework established a link between science, sustainable development, energy development, transportation, industrial development and pollution. Most importantly, it set a framework to reduce greenhouse gas emissions to 'prevent dangerous anthropogenic interference with the climate systems' (Jackson, 2007). In 1997, the UNFCCC was operationalised through the Kyoto Protocol where industrial countries committed to limit and reduce their greenhouse gas emissions in accordance with individually set targets (UNFCCC, 1997). The Protocol ensured that countries that have emitted the most greenhouse gas emissions will be the first ones to adopt policies and measures to reduce their emissions.

Since then, yearly negotiations and conferences have taken place where countries can update their commitments for climate action. The Paris Agreement is the most significant internationally binding treaty adopted in 2015 at the UN Climate Change Conference (COP21) by 195 member countries. For the first time, countries committed to the goal of limiting global warming to a set temperature. The member states committed to:

Holding the increase in the global average temperature to well below 2°C above pre-industrial levels and pursuing efforts to limit the temperature increase to 1.5°C above pre-industrial levels, recognising that this would significantly reduce the risks and impacts of climate change (UNFCCC, 2015:3).

Out of all the Parties in the UNFCCC, only three countries have not ratified the Paris Agreement, namely Iran, Libya and Yemen (UN, 2023a). The agreement aims to rapidly reduce emissions and to reach net-zero by 2050. Net-zero signifies that there is a balance between the greenhouse gases emitted and the emissions removed from the atmosphere by sinks that absorb the emissions. The sinks can be natural, such as forests that absorb CO_2, or be made through technological processes, such as carbon capture and storage. The Agreement includes requests for the countries to commit to pledges of emission reduction. The countries submit Nationally Determined Contributions (NDCs) which should be updated at least every five years. However, there are no guidelines on how countries should reduce their emissions. The only requirement is that the NDCs should have a higher degree of ambition in each new round of submission. While it is mandatory to provide NDCs, they are in themselves non-binding. Countries can also provide long-term low greenhouse gas emissions development strategies (LT-LEDS) to showcase how they plan to reduce emissions. The LT-LEDS are not mandatory and have only been submitted by 68 countries (UNFCCC, 2023a). The Paris Agreement highlights that technological innovation and fully realising technology development by adopting the best available technology is an important strategy to reduce emissions. In addition, the agreement underlines that the reduction of emissions should be based on the principles of equity and sustainable development. For instance, a requirement is to provide financial and technical capacity building for countries that are in need of support to adapt and mitigate climate change.

Despite committing to reducing emissions since the 1970s, and agreeing on a set temperature limit since 2015, the world is still experiencing a continuous increase in greenhouse gas emissions and increasing temperatures (WMO, 2023a). In fact, since the first report by the IPCC published in the 1990s, more CO_2 has been emitted than throughout all of human history. At the same time, the years following the Paris Agreement have consecutively been the warmest on record (WMO, 2023a). The rate of warming has more than doubled in the previous 40 years compared to the period before 1980 (NCEI, 2023). Currently, the global annual mean temperature is around 1.2°C warmer than pre-industrial temperature only 0.3°C away from the aspirational limit set in the Paris Agreement (WMO, 2023a). Now, the IPCC warns that the world is on track to reach 1.5°C annual global mean warming between 2030 and 2052 (IPCC, 2018). The most recent estimation is that it will be reached in the early 2030s (IPCC, 2021).

The UN Environmental Programme predicts that countries' NDC pledges made by 2023 would lead to a 2.5–2.9°C temperature rise above pre-industrial levels by 2100, highlighting that countries still fail to cut emissions sufficiently (UNEP, 2023a). This is alarming because a warmer planet brings increasingly devastating effects on our societies and ecosystems. Although the aspirational limit in the Paris Agreement is set to 1.5°C, an even lower temperature rise is preferable. Already at a 1.5°C warming, around 14 per cent of the world's population will be likely to face extreme heatwaves at least once every five years (NASA, 2019a). Extreme heat days can temporarily reach temperatures 3°C warmer than pre-industrial times. Additionally, it can lead to almost one metre of rising seas, extensive loss of habitat and biodiversity and 1.5 million tonnes of fishery loss (IPCC, 2018). A 1.5°C warming will also trigger accelerating glacial melt and thawing permafrost which in turn will increase carbon release further. The consequences of a 2°C warming would be even more extreme. Then, at least 37 per cent of the population will likely face extreme heat waves at least once every five years (NASA, 2019a). A 3°C warming would be catastrophic with the transgression of several vital tipping points that would leave us with no return to stabilised Earth systems (UNEP, 2023a).

While the international negotiations to meet the challenges of climate change have brought some hope of collective action, they are far away from being sufficient. Global warming has already resulted in weather extremes, humanitarian crises, food and water insecurity and forcing large population displacement. Climate change has already seriously threatened peace and security at various levels. Climate-induced natural disasters will continue to be worse as the world fails to find ways to mitigate and adapt to this serious challenge, which has the potential to lead to more political instability and create more conflicts.

PRIVATE SECTORS AND SOCIAL MOVEMENTS

The vast scale of the climate change challenge indicates that decisive collective action is inevitable. In this challenge for survival, nation-states are the primary actors. Nation-states have the political decision-making power and popular mandate for decisive action that is necessary to reduce emissions. The private sector also has a role to play in climate action, but mainly supportive. Several global companies have recently committed large sums of money to help reverse the impact of climate change. For instance, the founder of Patagonia, an American retailer of outdoor recreating clothing, gave up the ownership of the US$3 billion company in 2022 to fight climate change (McCormick, 2022). Now, the company is instead owned by a climate-focused trust that aims to redistribute money to protect the planet after money has been reinvested in the company (Patagonia, 2023). The company is expected to contribute US$100 million annually to fight the climate crisis (Porterfield, 2022). Although the initiative by Patagonia is quite unique, several companies have in recent years committed large amounts of money to contribute to fighting climate change. However, companies' investments in climate action strategies have many times been called out as greenwashing. Greenwashing is a strategy to portray a company, its products or its services as environmentally friendly, while they in fact contribute to environmental destruction or climate change. In other words, it can be considered false marketing that mislead the customers.

Many of the global leading companies are largely responsible for emitting greenhouse gases. In 2017, a report by the Carbon Disclosure Project revealed that only 100 companies worldwide are accountable for more than 70 per cent of the world's greenhouse gas emissions (CDP, 2017). Undoubtedly, the private sector is an important actor in limiting global warming. However, to ensure real commitment to substantially reduce emissions and environmental degradation, nation-states must steer by enforcing decisive climate action policies. In other words, given the vast challenges connected with climate change, the climate crisis is not something that can be adequately addressed by some well-meaning business leaders. They can, at best, play a supporting role, while the primary actors in this unprecedented challenge for survival are the nation-states. Unless the political leaders in power take the climate crisis seriously and prioritise fighting it, there is no way the world can survive the danger of global warming, sea-level rise and the increasing number of devastating natural hazards. Although there have been occasional encouraging public speeches, numerous international negotiations and the implementation of some climate action strategies, countries are mainly busy blaming others for climate change instead of taking concrete steps to limit it.

Nation-states have a crucial role in ensuring that the fight against the climate crisis is just and equitable. Currently, the richest one per cent of the world's population is accountable for producing as much CO_2 as the poorest two-thirds of the population, around five billion people (Oxfam and SEI, 2023). At the same time, it is generally the poorest parts of the global population that are most vulnerable towards the consequences of climate change (ND-GAIN, 2023). This is a serious challenge because, since 2020, the world's five richest men have doubled their fortune, while the poorest five billion people are becoming poorer (Oxfam, 2024). Unfortunately, there is a huge gap between the international commitments on human security and climate change and the national policies adopted by the countries. Climate change and national policy responses to meet its challenges will have a significant impact on the human security of millions of people. Only comprehensive and collaborative actions by nation-states in line with protecting human security will make it possible for the planet to meet these unprecedented challenges. Countries must commit to ambitious climate mitigation targets to keep the global average temperature increase within a manageable limit. Countries providing climate mitigation assistance and those who are receiving this support must commit to protecting human security. They must incorporate human security norms into their domestic legal framework. While countries need to take important steps towards fulfiling their obligations at home, they need to work cooperatively with other countries to combat climate change and ensure the protection and well-being of people across the world.

In light of this, the world is also witnessing growing climate justice movements. NGOs and youth movements are calling for emissions reduction and climate justice worldwide. While these movements can bring hope and engagement to the public, they cannot solve the climate challenges by themselves. Unless the climate change issue becomes countries' top priority, whatever some well-meaning individuals, business houses and civil society groups do will not be enough to save the planet. While many interest groups and social movements project climate change as a threat to human security, there is an overall reluctance to elevate that threat to the level of national security. The reluctance is primarily based on the fear that in making climate crisis a national security issue, countries will secure their borders and territories, and ignore the aspects affecting human security. There are many ways climate change poses serious threats to national security, and those need to be highlighted in national discourse. Highlighting the impact of climate change on national security is not denying the importance of its threat to human security but facilitating the process for the political leaders of the countries to get the popular mandate to take urgent and concrete measures towards mitigation and adaptation to global warming.

SUMMING-UP

In the evolving world of security, the term has transformed from a narrow focus on military might and state sovereignty to a multifaceted concept shaped by global dynamics. In the past, security was all about guarding borders and deterring military threats, especially during the high-stress Cold War era. Fast forward to the post-Cold War period, and we see a dramatic shift: the world now recognises a spectrum of security challenges extending beyond traditional military concerns. This new era of security acknowledges the impacts of civil wars, transnational crimes, terrorism, cyber wars, hybrid wars and infectious diseases. Globalisation has played a pivotal role in this evolution, expanding the security narrative to include economic crises, societal disruptions and environmental challenges. This broad perspective recognises that security isn't just about protecting state territories; it's about the well-being of societies and individuals, understanding that global challenges are interconnected and have ripple effects across borders.

The security discourse is now a vibrant debate between traditionalists, who focus on military and state sovereignty, and wideners, who advocate for a more inclusive approach recognising various threats, including non-military ones. This debate has given rise to the concept of human security, which centres on the individual. Introduced by the UNDP in 1994, human security shifts the focus from armament to sustainable human development, highlighting the protection of human rights, safety and survival. Defining security in the modern world is complex, requiring a multifaceted approach that balances diverse interests. This complexity demands an adaptable strategy, considering not just military or territorial threats but also factors contributing to societal peace and security. The current security landscape is diverse, encompassing challenges like hybrid wars, cyberattacks and identity-based civil wars. These issues call for a comprehensive security architecture that extends beyond traditional military deterrence, focusing on multilateralism and smart resource management.

Undoubtedly, security in the 21st century transcends traditional military power. It's about an integrated approach that addresses global, national and human security, ensuring peace through collaborative efforts that address economic, societal, cultural and environmental challenges. In recent decades, climate change has emerged as a non-military security challenge, recognised as a critical threat to global stability. Human activities since the Industrial Revolution have led to a rapid rise in global temperatures, creating and exacerbating security challenges. The effects are far-reaching: heatwaves, wildfires, changes in weather patterns, droughts and flooding. These changes not only impact the environment but also have profound societal consequences, affecting water and food security, damaging homes and livelihoods and causing population displacement. Moreover, climate change also leads to feedback loops, exacerbating global warming and risking irreversible changes in environmental systems. Economically, it has already caused significant losses, and it poses geopolitical risks, especially in regions like the Arctic, the Middle East, and North Africa, which are experiencing faster warming. This uneven impact triggers large-scale migration, fuels tensions and conflicts and affects economic activities.

Climate security, therefore, has evolved in response to the growing recognition of climate change's impacts on global stability. It has expanded the traditional understanding of security to include environmental issues. The early 21st century marked a significant development in climate security, with mounting evidence of climate change's potential impacts on global stability. This has led to the gradual integration of climate considerations into national and international security policies. Non-state actors like NGOs, academia and the private sector have played crucial roles in advancing the climate security agenda. Climate security intersects with global issues like human rights, development and migration. It recognises climate change as a 'threat multiplier', exacerbating existing vulnerabilities and heightening conflict risks. Despite international efforts, global greenhouse gas emissions and temperatures continue to rise. The private sector's contributions, while important, are insufficient. The responsibility for combating climate change primarily falls on nation-states, which must enforce decisive climate policies and ensure equitable solutions. Additionally, there's a growing movement for climate justice, driven by NGOs and youth groups. These movements, while raising awareness, cannot solve the climate crisis alone. Nation-states must prioritise climate change as a top policy issue and take concrete steps towards mitigation and adaptation. Highlighting the impact of climate change on national security is essential for urgent and comprehensive climate action in a collaborative manner at the global level.

2
CLIMATE APATHY AND SELF-PRESERVATION

This chapter delves into the profound yet perplexing apathy exhibited by political leaders in addressing the climate crisis – a global challenge that has, paradoxically, become both a politically divisive and politically costly issue. At the heart of this dissonance lie the inherent limitations of democratic institutions, marked by short-term thinking and a propensity for compromises that often dilute the urgency and magnitude of environmental threats. While revisiting the inadequacy of political responses, it probes the extent to which political leaders and institutions are failing to mirror the criticality of the situation. This chapter also examines the protectionist stances adopted during international negotiations, reflecting a disheartening prioritisation of immediate national interests over long-term global survival. Further, it tries to demonstrate the dichotomy between the undeniable reality of climate risks and the hesitation of political figures to take decisive action, despite increasing public awareness and demand for change. The influence of state and private sector funding in spreading misinformation adds another layer of complexity to this scenario, clouding public perception and hindering progress. Moreover, the impact of lobbying on fostering protectionist, incrementalist and status quo approaches is scrutinised, revealing the intricate web of interests that impedes transformative policies. Despite these challenges, this chapter acknowledges some progress in the realm of multilateral and bilateral agreements, offering a glimmer of hope in an otherwise grim landscape of political inertia and short-sightedness.

POLITICAL RESPONSES TO THE CLIMATE CRISIS ARE INADEQUATE

The climate crisis poses a severe threat to our societies. Current political responses are insufficient to effectively address this crisis. Primarily, political apathy and self-preservation are impeding urgent climate actions required to prevent extreme and catastrophic outcomes. While political leaders convene annually to discuss strategies for tackling climate change, existing commitments fall short of reducing emissions and ensuring sustainable societies for all. The Paris Agreement, adopted in 2015, represents one of the most significant international political responses to climate change. For the first time, the world agreed to limit global warming to 'well below' 2°C compared to pre-industrial (1850–1900) levels, with an aspirational target of 1.5°C. The continuously increasing temperature and emission levels show the failure of political responses established in the last five decades.

Although the current long-term global mean temperature is 1.2°C above pre-industrial levels, the global annual average for 2023 temporarily exceeded a 1.5°C warming (Rhode, 2024). In 2023, June, July and August were particularly warm, accounting for the hottest months on record, with a monthly global mean warming in July and August of 1.5°C. This marked the first time that summer months and an entire year reached a 1.5°C warming. Simultaneously, the oceans are warming at an unprecedented pace. The extreme heat observed in Europe, North America and China during the summer of 2023 significantly contributed to this temporary warming. In fact, without anthropogenic climate change, the extreme heat experienced during this period in these regions would have been nearly impossible (Zacharias et al., 2023).

The ongoing and increasing global greenhouse gas emissions are driving planetary warming. The Intergovernmental Panel on Climate Change (IPCC) estimates that global greenhouse gas emissions must peak before 2025 to have a 50 per cent chance of limiting warming to 1.5°C (IPCC, 2023a). Consequently, the world must urgently commit to climate action to promptly reduce emissions. However, greenhouse gas emissions are continuing to rise rapidly. In 2022, the world witnessed the highest level of greenhouse gas emissions ever recorded. The primary contributor to this surge in carbon emissions is the increasing use of coal and the continued burning of fossil fuels (IEA, 2022a). While there was a slight reduction in global CO_2 emissions during the COVID-19 pandemic, emissions in 2022 surpassed pre-pandemic levels, indicating a continued upward trajectory (WMO, 2022). Atmospheric methane levels, a potent greenhouse gas, are also increasing. Methane is 25 per cent more effective at trapping heat than CO_2 (WMO, 2023a). Agriculture and the energy sector, including oil, coal, natural gas and biofuels, are the main sources of anthropogenic methane emissions (IEA, 2021a). Despite more than 100 countries pledging to reduce their methane emissions by 30 per cent compared to 2020 levels by 2030, emissions are rising rapidly (Global Methane Pledge, 2021). In fact, between 2020 and 2021, the world experienced the largest annual increase in methane emissions on record (WMO, 2023a). Given the continued rise in emissions, the target of significantly reducing greenhouse gas emissions to limit global warming remains elusive. In essence, political responses to address the climate crisis have thus far proven inadequate.

The inadequacy of political responses to emissions reduction can be partly attributed to the voluntary nature of the Paris Agreement. The Agreement only requires countries to voluntarily pledge their Nationally Determined Contributions (NDCs) to reduce greenhouse gas emissions. Pledges to cut emissions do not necessarily translate into concrete actions to fulfil these promises. There are no formal sanctions for countries that fail to follow through or make insufficient pledges. In addition, the Agreement itself does not explicitly address one of the primary contributors to climate change, namely fossil fuels. Despite fossil fuels accounting for 75 per cent of global greenhouse gas emissions and 90 per cent of CO_2 emissions (UNFCCC, 2015), the Agreement does not explicitly mention them. Similarly, annual international climate negotiations in the Conference of the Parties (COP) have for a long time failed to adequately address the root causes of climate change. Despite the growing urgency of the climate crisis, recent COP negotiations have made little progress in transitioning away from fossil fuel usage, despite the evidence that supports this necessity.

One notable manifestation of the failure to prioritise climate change in international affairs is the absence of significant world leaders at climate negotiations. Even top polluters like the United States withdrew from the Paris Agreement under the Trump administration, although it subsequently rejoined in 2021 under the Biden administration. However, world leaders often make brief appearances at climate conferences and do not fully engage with the pressing issues. For example, at COP27 in 2022, President Biden's stay was relatively short before departing for another meeting (Liptak and Nilsen, 2022). At the 2023 COP28 in Dubai, Joe Biden and Xi Jinping did not even attend, despite their countries being major polluters. This absence dampened hopes for a productive outcome from the conference. Consequently, climate negotiations continue year after year, discussing concepts such as climate funds, carbon credits, and loss-and-damage funding. However, these talks frequently result in deadlock, evasion of key issues or limited progress for real change. The final document agreed on at the 2023 COP28 in Dubai has been referred to as the 'beginning of the end of the fossil fuel era' because of the inclusion of the aim to 'transitioning away from fossil fuels in energy systems in a just, orderly, and equitable manner' (UNFCCC, 2023b:5; UNFCCC, 2023c). However, there is no reference to oil or gas in the document (UNFCCC, 2023b). In addition, transitioning away from fossil fuels does not necessarily mean phasing out fossil fuels. Instead, it leaves room for interpretations of what transitioning away implies.

THE URGENCY OF THE CLIMATE CRISIS IS GOING UNRECOGNISED

Although there has been some minor progress in climate negotiations, there is a noticeable lack of meaningful action to address the root causes of climate change, particularly the phasing out of fossil fuels. The longstanding reluctance to address fossil fuel reduction at the international level translates into climate inaction at the national level. Governments worldwide continue to plan for increased fossil fuel production, despite ambitious climate targets and net-zero commitments. The planned fossil fuel production for 2030 is projected to be double what would be consistent with limiting warming to 1.5°C (SEI et al., 2021). A detailed analysis of countries' NDCs and Long-term Low Emissions and Development Strategies (LT-LEDS) submitted up to March 2023 reveals that only a few countries mention the need to reduce fossil fuel production (Jones et al., 2023). In the first round of NDCs, only India and Nigeria addressed the necessity of reducing fossil fuel production to meet their climate goals. In the second round, Costa Rica, North Macedonia, Pakistan and EU countries mentioned policy targets for reducing fossil fuel production. However, expressing interest in reducing production is not translated into active policymaking in these countries. Instead, the reduction of production is framed as a potential future outcome resulting from reduced demand by some countries. For some, the reduced demand is even perceived as a national risk due to economic dependence on these resources. For instance, Norway highlights the expected reduction in fossil fuel demand as a 'transition risk' to its economy (Jones et al., 2023:15). Moreover, one-third of the submitted second-round NDCs do not even mention fossil fuel production, as it is not a requirement under the Paris Agreement. Among the countries that do mention fossil fuels, fifteen explicitly express intentions to continue or increase their production (e.g., Canada, the United Arab Emirates, Saudi Arabia, the United States, and China). Countries use various arguments for the need for continued production, such as the need to protect national energy security, reduce energy costs or pursue economic growth. According to the NDCs, some countries even view continued fossil fuel production as a part of their climate mitigation efforts, by making the production 'greener' through more efficient technology. However, these plans to continue or increase production do not align with the countries' climate targets, particularly achieving net-zero emissions (Jones et al., 2023). The International Energy Agency emphasises that to have a chance of reaching net-zero emissions by 2050, the world must halt all oil, gas and coal development immediately (IEA, 2021b). The failure to prioritise a substantial reduction in fossil fuel production highlights the political shortcomings in addressing the urgency of the climate crisis.

For instance, in response to rising energy costs and supply disruptions caused by Russia's military invasion of Ukraine, President Biden released 30 million barrels of oil from the US Strategic Reserve. He also urged allies to release an additional 30 million barrels (DoE, 2022). These geopolitical tensions have shifted priorities away from addressing climate change. It was expected that, with the easing of the COVID-19 pandemic, the world would refocus on climate change. However, conflicts like the war in Ukraine and increased global instability have diverted attention away from the climate crisis. In times of war or fear of war, climate change often takes a backseat as a policy issue and an electoral concern. Instead, countries focus on mitigating energy security risks and inflation, which has been making non-Russian fossil fuels producers increasingly requested. Mainly, for EU countries to reduce their dependency on Russian fossil fuels. For instance, European fossil fuel producers like the United Kingdom, Denmark and Norway are now in high demand to meet short-term energy needs (Sanchez et al., 2023). However, in the long-term, the hope is that the war will accelerate the transition to renewable energy sources to reduce dependence on Russian oil and gas. Nevertheless, EU countries must urgently accelerate their emission reduction efforts to meet their ambitious goals. Based on their 2021 emission reduction rate, they would not achieve their target of reducing emissions by 55 per cent until 2051 (Enel and The European House Ambrosetti, 2021). At this rate, the European Union would reach its goal 21 years later than the planned year of 2030. The ongoing geopolitical tensions have significantly complicated the task of substantially reducing global emissions, emphasising the need for global leaders to prioritise climate change as a top global issue.

In conclusion, the climate crisis poses an existential threat to our societies, and current political responses have proven inadequate in addressing this crisis. Factors such as political apathy, self-preservation and geopolitical tensions have hindered the urgent climate actions required to prevent extreme and catastrophic consequences. Despite international agreements like the Paris Agreement, the world is still not on track to achieve critical climate targets. Greenhouse gas emissions continue to rise rapidly, primarily driven by the use of fossil fuels. The voluntary nature of international agreements and the lack of substantial action to phase out fossil fuels have contributed to this problem. On top of that, the absence of significant world leaders at climate negotiations and the diversion of attention to geopolitical conflicts have taken the focus away from the urgent climate crisis. The failure to prioritise substantial reductions in fossil fuel production, as well as conflicting actions and statements from political leaders, further highlights the inadequacy of current political responses. Urgent action is needed at both the international and national levels to address the climate crisis effectively and safeguard our planet's future.

PROTECTIONISM

While the climate crisis can pose threats to countries' national security, a single country cannot solve the climate crisis alone. It requires nothing less than global collective action. The realisation of this has convinced countries to take steps towards collectively acting on reducing global warming. Why are the current efforts so notably insufficient? A main cause of this is countries' inability to reach a consensus on necessary actions and the reluctance to respond urgently. Global action demands that all nations step up their efforts to reduce their emissions. All countries would benefit from significantly limiting global warming. However, climate change is often depicted as a future challenge, with hope pinned on future solutions. Drastically reducing national emissions can be seen as potentially weakening a state's economy or military power in the short-term, as both sectors might be heavily dependent on fossil fuels. National incentives to reduce emissions can, therefore, be low, especially if other countries are not transitioning at the same speed. This difficulty is exacerbated if a country's economy relies on fossil fuel production or other emission-intensive industries. Consequently, it may seem that each country would benefit the most if other countries were more accountable for large emission reductions and associated costs.

Protectionist approaches, aimed at safeguarding national security and interests, significantly contribute to the failure of global climate negotiations. These approaches are often characterised by the imposition of tariffs and other trade barriers, which hinder the international flow of green technologies. As a consequence, the global adoption of renewable energy sources and energy-efficient products is slowed down, with these technologies becoming more expensive and less accessible beyond their countries of origin. Such protectionism also leads to a reluctance among nations to commit to, or comply with, international climate agreements. Countries focused on their immediate economic interests may find it challenging to enact policies crucial for global climate mitigation, especially when these policies might seem detrimental to their own industries. Moreover, while these protectionist measures might appear to serve short-term national interests, they present considerable obstacles to the collective action needed for effective climate change mitigation. Addressing global challenges like climate change necessitates cooperative strategies and policies that go beyond national boundaries, emphasising the need for a more collaborative and less protectionist approach in the international arena.

The blame game between countries is significantly hindering progress in achieving meaningful global climate action. It often revolves around assigning responsibility for climate change and, consequently, who should take immediate and drastic measures to reduce their emissions. For example, China currently stands as the largest emitter of greenhouse gases, accounting for over 30 per cent of the world's greenhouse gas emissions (World Bank, 2022a). Alongside the United States and India, they contribute to nearly 43 per cent of global emissions. Globally, only ten countries are responsible for 60 per cent of all greenhouse gas emissions (see Figure 2.1). In contrast, the 100 least-emitting countries produce less than 3 per cent of global emissions (Climate Watch, 2023). China, as the world's largest emitter, often faces blame for not doing enough to kerb its emissions. As a result, some countries with high emissions justify their inaction by placing the bulk of the responsibility on China to rapidly reduce emissions. However, considering China's large population, it doesn't even rank among the top 40 emitters on a per capita basis. From this perspective, countries like Qatar shoulder a larger share (Ovaska et al., 2021). Typically, countries with small populations and emission-intensive industries rank among the top countries in emissions per capita (see Figure 2.2). Among the top ten emitters globally, the United States and Russia have the highest per capita emissions (Vigna and Friedrich, 2023). India, despite being the third-largest emitter globally, falls to 131st place when considering emissions per capita, owing to its large population (Ovaska et al., 2021).

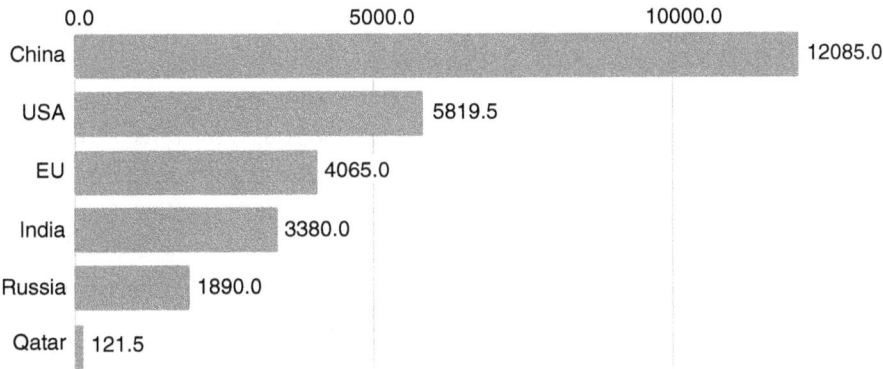

Figure 2.1 Emissions per Country

Note: The graph shows the top emitting countries total greenhouse gas emissions in MtCO$_2$e. The graph also shows the total emissions in Qatar, the country with highest per capita emissions.

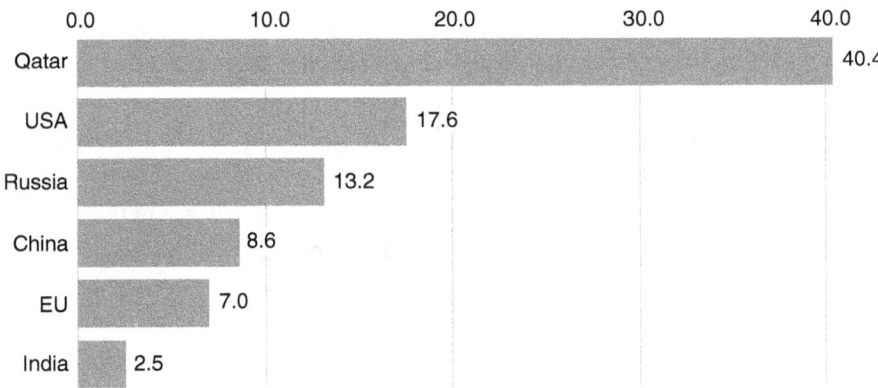

Figure 2.2 Emissions per Capita

Note: The graph shows the emissions per capita in tCO$_2$e/person, comparing Qatar, the highest emitting country per capita globally, to the top total emission countries globally.

Another perspective considers historical emissions, dating back to the Industrial Revolution. From this viewpoint, the United States and European Union emerge as the largest cumulative emitters by a significant margin. For instance, the cumulative emissions from the United States are more than twice those of China (Evans, 2021). The Industrial Revolution and its cumulative emissions have led the United States and European Union to become wealthy high-income countries and are often held responsible for initiating the climate crisis. Hence, they are considered to bear more significant responsibility for rapidly reducing emissions, given their affluence. Especially compared to countries with lower socio-economic development that have historically made minimal contributions to climate change. Recent global climate agreements reflect this perspective by emphasising that developed countries should take the lead in reducing emissions and providing financial aid to less affluent but more climate-vulnerable nations (UNFCCC, 2015). However, it's not uncommon for Western countries to blame countries like India and China for their large total emissions, without considering the cumulative effects and emissions per capita. For instance, in 2017, then-President Trump argued that the Paris Agreement would hinder his administration's ability to shape domestic environmental laws according to the country's needs, as it would prevent him from adhering to his 'America first' foreign policy. Consequently, he decided to withdraw American contributions to climate finance, asserting that the Paris Agreement would only marginally impact global temperatures and wasn't strict enough on countries like India and China (Trump White House, 2017). In other words, these varying perspectives on responsibility and accountability for the climate crisis fuel the blame game and complicate the process of reaching a consensus on global climate actions. Ultimately, emissions reduction is a collective responsibility that necessitates global actors to cooperate and coordinate rather than protect and promote their own national interests. Therefore, national security in the context of climate change should not be within a nationalist framework but given the severity of the issues at hand, countries should open them up for cooperating with others.

Unfortunately, the pursuit of national interests often delays urgent collective climate action. For example, India and China successfully pushed for a change in the language from a 'phase out' to a 'phase down' of unabated coal power in the final document of COP27 in Glasgow. This language shift lowered the ambition of the overarching global climate target while aligning with the two countries' own agendas (Frey and Burgess, 2022). After the COP28 discussions in Dubai, unabated coal power is still referred to as needing to be phased down, rather than phased out (UNFCCC, 2023b). Consequently, they are now under less pressure to entirely phase out coal use, despite coal consumption being a major contributor to climate change. Remarkably, nearly 95 per cent of global coal consumption occurs in countries committed to achieving net-zero emissions (IEA, 2022b). India, for instance, has pledged to attain net-zero emissions by 2070, reduce carbon intensity significantly by 2030, and accelerate the transition to renewable energy sources. However, India's national energy plans conflict with its climate targets, as it plans to significantly expand coal power capacities between 2027 and 2032 (Climate Action Tracker, 2023a). This misalignment between India's rhetoric and actions casts doubt on its ability to transition to clean energy as outlined in the Paris Agreement.

China, despite previously planning to increase coal production, has recently strengthened its climate commitments domestically. It now intends to gradually reduce coal production starting in 2025 (Climate Action Tracker, 2023b). Additionally, China is rapidly developing renewable energy sources and boasts the world's largest electric vehicle fleet and battery market (World Bank, 2022a). However, while China is taking more responsibility at home, it is exploiting the environment outside its borders to sustain its growing economy and competitive global position. For example, China, along with India and other countries, has reduced domestic forest logging while increasing imports of forest products. This has led to deforestation in other countries, particularly in Laos and the Indochina region, thus contributing to increased greenhouse gas emissions. Moreover, in 2017, China abandoned plans for large hydropower dams upstream of the Nu-Salween River due to environmental concerns but pressured Myanmar to permit Chinese companies to construct a series of downstream hydropower dams for Chinese consumption (Fawthrop, 2017). China's Belt and Road Initiative, which involves extensive international infrastructure projects, has also faced accusations of environmentally damaging energy projects in other countries, along with accusations of burdening developing nations with substantial debt. In essence, while China is taking concrete steps to reduce greenhouse gas emissions within its borders, its economic power allows it to export environmental degradation to other countries.

Outsourcing emissions and environmental harm are not unique to China. For instance, the EU population contributes to environmental damage and high emissions levels by heavily relying on imports of commodities. While the continent has reduced emissions per capita by 29 per cent between 1990 and 2019 through the adoption of renewable energy and energy-efficient technologies (Vigna and Friedrich, 2023), European citizens import approximately one ton of CO_2 per person annually in products from outside Europe (Fuchs et al., 2020). The United States stands as the world's largest importer of emissions, importing 131.5 million metric tonnes of CO_2 emissions annually. In this context, China, often referred to as 'the world's factory', is the largest exporter of emissions. Inevitably, national commitments may fall short if they fail to consider how their actions impact emissions and environmental degradation in other countries.

In light of this, global collective action and cooperation are imperative to limit global warming. Notably, cooperation between the United States and China, as the world's largest emitters, plays a pivotal role. Their cooperation is essential not only because of their enormous contributions to global emissions but also for their leadership roles in the world and their capacity to drive technological innovation and market transformations. By working together on climate action, they can significantly advance the global effort to limit global warming and mitigate its adverse effects. Without effective collaboration between these two major powers, the world is likely to fall short in addressing the climate crisis.

WHY POLITICIANS FAIL TO TAKE ACTION

Despite inadequate political responses to the climate crisis, the implications of climate change are well known to political leaders. The 2024 Global Risk Report finds that over the next 10 years, the planet's health dominates global risk perceptions. The 2023 Global Risk Report had similar results, with the failure of climate action to mitigate and adapt to the changing climate ranked as the highest risk (WEF, 2023b). In other words, global risk perceptions and numerous climate negotiations indicate that the severe implications of climate change should be well known to political leaders. Generally, political leaders can respond to climate risks by resisting and avoiding the issue, implementing adaptive measures or using the debate to fulfil their own agenda (Earle et al., 2015).

Globally, the majority of the population is concerned about climate change. The World Risk Poll of 2021 finds that while 41 per cent view climate change as a 'very serious' threat, 26 per cent perceive climate change as a 'somewhat' serious threat. On the contrary, only 14 per cent don't find climate change a threat at all (LRF, 2022). Even in the United States, there is a major shift in public perception about climate change threats. Though climate change is somewhat of a polarised issue, the number of people viewing climate change as an urgent threat is increasing. In fact, during the latest decade, the number of people being 'alarmed' by global warming in the United States has increased substantially, whereas the dismissive section of the population is losing its numbers. In 2012, 12 per cent of the US population was alarmed by global warming, a number that has grown to 26 per cent in 2022. In general, the Yale Program on Climate Change Communication finds that US citizens are becoming more worried about climate change and becoming more engaged in dealing with the issue (YPCCC, 2023).

There is an increasing recognition of climate risks among voters, and they expect their governments to act. For the 2022 US Congress vote, almost 60 per cent prefer voting for candidates who are willing to support climate actions. Meanwhile, only 14 per cent of the registered voters believe that the United States is responding well to the climate crisis (YPCCC, 2022). However, even though many electorates are concerned about climate change and want to see concrete action, climate change issues are rarely placed as top priorities in electoral processes. There seems to be a mismatch between what voters want and what they prioritise. For instance, for the 2022 US Congress election, climate change was ranked the 24th most important issue for voters (YPCCC, 2022). Instead, the economy, education, and health care were top priorities among voters (Schaeffer and van Green, 2022). A similar lack of prioritising climate change in elections is happening in many countries. For instance, during the Spanish election, amidst extreme heat temperatures and wildfires in July 2023, climate change was barely discussed by the campaigning politicians. Only one per cent of the Spanish population addressed climate change as the most urgent issue according to pre-election polls, whereas the economic situation was considered the most urgent issue by the majority (Kuper, 2023). Also, during the 2022 election in France, climate change was almost completely absent in the political debates and speeches. Although the public ranked environmental challenges, including climate change, as the third most important topic according to pre-election polls, only around 1.5-5.5 per cent of campaigns covered by media discussed climate change (Garric, 2022).

The EU country with the most climate-concerned population is Sweden (EU, 2021). In Sweden, around 70 per cent of the population is worried about climate change (Naturvårdsverket, 2021). The pre-election polls for the 2022 election indicated that only around eight per cent of the population does not care about climate issues (PwC, 2022). Even though climate change and other environmental challenges were discussed by the politicians in the pre-election debates, they mainly covered issues related to energy. Predominantly, the discussions focused on electricity production, the increase or phase down of nuclear power production and how to tackle the rising electricity prices (Miljand, 2022). Regrettably, discussions about other kinds of climate mitigation and adaptation strategies were almost completely absent. Conversely, in the 2021 election in Norway, climate change dominated the political discussions. Climate change was one of the top priorities for voters according to a pre-election poll (Aftenposten, 2021). The election was even referred to as the 'climate election'. Despite this, parties with ambitious climate approaches, such as the Green Party, fared less well than expected. The Green Party did not even reach the electoral threshold of four per cent (Farstad and Aasen, 2022). One of the major topics discussed in relation to climate change was the production of Norwegian oil, as the country is one of the biggest oil producers in the world. The Green Party was the only Party that had a set limit for continued Norwegian oil production, in 2035 (SVT, 2021). The winning parties instead opted for continued exploration and production with extra risk assessment and some reservations. In other words, even though climate change was one of the main topics discussed during the election, the result showcases unwillingness for decisive action to drastically reduce emissions.

As climate change has become a partisan issue, political leaders are becoming unwilling to take decisions carrying electoral risks and political costs. Any forceful action or policy can cost electoral votes. It is well known that long-term costs for unrestrained global warming will outweigh many times the short-term adjustment cost for climate mitigation. However, it is a fact that political leaders are bound by the practicalities and procedures of everyday politics and, in some cases, regular election cycles. There is a need for strong and wise political leadership to agree for their country to bear the immediate costs on an equitable and credible basis. Even if a consensus is reached in international climate negotiations, measures might not be implemented on a national level due to the political challenges it may provoke. Mainly, the rise of right-wing populist leaders in some key countries has shown setbacks to global cooperation against climate change (Frey and Burgess, 2022). Climate change can become a partisan issue because while some argue that politicians are doing far enough for the climate crisis, others believe that climate science is unreliable and that politicians exploit climate change for political gain (Meijers et al., 2022).

In the United States, Biden was elected to the White House by promising his supporters to be the 'Climate President' and has pledged that the country will reach net-zero by 2050. During the election campaigns, he promised to end the reliance on oil and gas to fight climate change (BBC, 2020). During his presidency, he cancelled the lease of a million acres of land that was given to oil and gas companies for possible drilling (Phillips, 2022a). He has cancelled oil and gas leases in Alaska that were issued under Trump's presidency to protect the irreplaceable Alaskan wildlands. In addition, he revoked a key permit and forced the cancelation of the high-profile Keystone XL pipeline that was expected to carry 830,000 barrels of oil sands per day from Alberta in Canada to Nebraska in the United States. The pipeline was opposed by Native Americans and environmentalists (Reuters, 2021). During the US midterm elections in 2022, he even accused oil companies of 'war profiteering' on oil prices while not lowering the price for the customers and threatened to impose a windfall tax (The Guardian, 2022).

At the same time, despite its ambitious climate targets, in 2023 the Biden Administration approved a new oil project in the Alaskan Northern Slope. The oil project is estimated to contain 600 million barrels of oil and generate up to US$17 billion in revenue (Reuters, 2023). The project, known as the Willow Project, is planned to consist of five drilling sites and produce 180,000 barrels of oil daily, equivalent to about 1.5 per cent of the total US oil production (Brown and Bohrer, 2023). Importantly, the Willow Project is expected to emit 263 million tonnes of greenhouse gases over its 30-year lifespan, equivalent to the total emissions from two million cars over the same period or the emissions of 76 coal plants running for a year (Osaka, 2023). This decision raises questions about the Biden administration's commitment to being a 'climate president' and the US's leadership on climate issues. The Willow Project is a huge setback to the Biden administration's climate policy record and upsets one of his core support bases, primarily youth and activists (Friedman, 2023). At the same time, the Willow project has widespread support from local communities in Alaska, as it will boost the state's economy and energy production (Nilsen, 2023). Campaigning for election is one thing, but the art of governing is another altogether. While campaigning, Biden had promised to end fossil fuel and become the 'Climate President', and he had also vowed to listen to local communities. Like in any other democracy, the American President realises the 'democratic dilemma' while deciding on the Willow Project.

An important takeaway from the decision-making process over the Willow Project is that democracies, even if it is the United States, can be hesitant to undertake the long-term problem-solving approach that climate change demands. It will try to find a compromise, and these short-term escape routes will not be enough to confront the survival crisis that climate threats pose to the planet. Political leaders and governments often operate within short-term election cycles, focusing on immediate concerns over long-term issues like climate change. Long-term environmental strategies may not align with short-term political goals. Politicians are in general hesitant to implement policies that might be unpopular or affect their electability. To address this, political leaders need to see the benefits of working in cooperation under a multilateral framework where the risks and costs could be lowered and shared. At this critical juncture, before it becomes too late, humankind needs political will and leadership, which can take their people together in this journey to securing the future. The threat of climate change is real, the world has already started to suffer, and the impacts are expected to intensify in the coming decades. Even school-going children are mobilising worldwide and asking adult policymakers to act like adults to protect their future. For instance, the 'Fridays for Future' movement, initiated by Swedish activist Greta Thunberg, has played a pivotal role in amplifying the urgency of addressing climate change. The movement has been instrumental in engaging young people worldwide, giving them a voice and platform to express their concerns about their future in the face of climate change. This engagement has led to a significant increase in awareness and activism among younger generations. The movement underscores the urgency of the climate crisis. The regularity of the protests, symbolised by the 'Friday' actions, serves as a constant reminder of the ongoing and pressing nature of the issue, mobilising people and resources towards addressing it.

Undoubtedly, the lack of political will has kept the world from taking concrete and decisive steps against climate change. There are significant economic interests tied to industries that contribute to climate change, like fossil fuels. These industries have substantial lobbying power and financial resources to influence political decisions, often hindering environmental policies. Moreover, achieving consensus on international policies and agreements can be a complex and slow process. At the same time, significant policy changes at home can lead to resistance from the public, especially if they involve economic sacrifices or major lifestyle changes.

MISINFORMATION AND PUBLIC FUNDS

Insufficient climate measures are not solely a result of political leaders' reluctance to take bold actions. The dissemination of misinformation regarding climate science is also a significant barrier to mounting effective responses to the crisis. False and inaccurate information about the climate crisis has been intentionally and unintentionally propagated by both private and public sectors for an extended period, aimed at discrediting climate science. In 2022, the IPCC recognised that misinformation about climate science has impeded urgent actions to combat climate change. This hindrance occurs primarily because such misinformation sows doubt regarding climate risks. Powerful economic and political interests are major drivers behind the spread of this misinformation (IPCC, 2022a). It's important to differentiate between misinformation, which is false or inaccurate information spread unintentionally, and disinformation, which is false information spread deliberately to deceive (Jack, 2017).

For example, the severity of climate change and its linkage to human activities has long been understood within the fossil fuel industry. This industry is responsible for substantial greenhouse gas emissions. Consequently, efforts to mitigate climate change can be perceived as a threat to fossil fuel-producing entities and nations. This perception arises from the potential compromise of their geopolitical, economic and military power if they can no longer rely on fossil fuel exports. To safeguard their interests, both private and public sectors have engaged in financing disinformation campaigns.

While large companies within the fossil fuel industry have been aware of the effects of climate change for decades, their public communications have often ignored climate change and its connection to human activities. Since the 1980s, there has been a clear association between the interests of the fossil fuel industry, climate deniers and climate-sceptical think tanks that disseminate misinformation. Over the years, major fossil fuel producers have funded conservative think tanks to produce disinformation on climate change. They have presented their own experts on the subject as unbiased authorities, dismissing climate science evidence by deliberately misinterpreting research findings. For instance, one of the largest fossil fuel-producing companies in the United States, ExxonMobil, has been aware of global warming since the 1970s (Supran and Oreskes, 2017). They employed their own scientists who generated predictive climate models in the 1980s and 1990s, which recent scrutiny has shown to be highly accurate. However, their public messaging contradicted these predictions (Supran et al., 2023). The goal has been to muddle the public's comprehension of climate change and create uncertainties (Union of Concerned Scientists, 2007). Fossil fuel producers have also played a substantial role in funding universities, thereby influencing climate science. Between 2010 and 2020, six fossil fuel producers funded climate and energy research at 27 US universities, totalling at least US$700 million

(Kumar, 2023). Additionally, the state-owned oil company Aramco in Saudi Arabia has invested approximately US$2.5 billion in US universities and funded almost 500 studies over the past five years. For instance, some of these studies aimed to question the effectiveness of electric vehicles (Tabuchi, 2022). Consequently, fossil fuel industries wield significant influence over the scientific and political discussions surrounding climate actions. The uncertainty surrounding the climate crisis, exacerbated by the spread of misinformation and disinformation, poses a significant barrier to taking urgent climate action. This issue is multifaceted, influencing both outright climate change denial and more subtly nuanced confusion 741 action. A clear example of this is the phenomenon of climate denial narratives, often fuelled by misinformation campaigns. These campaigns, sometimes backed by vested interests such as fossil fuel industries, disseminate false claims downplaying the severity or even the existence of climate change. A notable instance was the 'Climategate' scandal in 2009, where hackers leaked emails from climate scientists just before the Copenhagen Summit and misrepresented their contents to suggest scientific misconduct, thus feeding climate change scepticism (McKie, 2019).

Misinformation contributes to nuanced confusion, which can be equally damaging. An example of this is the over-reliance on future technological solutions to climate change. While technological innovation is crucial, the narrative that future technologies will solve all climate problems can lead to complacency in the present. For instance, the excessive emphasis on carbon capture and storage (CCS) technologies, which are still in their infancy and not yet deployed at a scale necessary to significantly mitigate climate impacts, can be misleading. This reliance can rationalise current slow or minimal climate action, under the mistaken belief that these technologies will efficiently and effectively resolve the issue in the future. In both cases, whether through overt denial or subtle delay tactics, the spread of misinformation and disinformation serves to impede the urgent and comprehensive action needed to address the climate crisis effectively.

Although the techno-optimistic narrative does not outright deny climate change, it defers to addressing the problem and solutions in the distant future. This kind of narrative is prevalent in many national and international climate agreements and policies. For example, Saudi Aramco, the world's largest state-owned oil-producing company, has set a climate target to achieve net-zero emissions by 2060. Simultaneously, Aramco plans to increase its oil production from 12 to 13 million barrels per day by 2027. To achieve its net-zero target, the country plans to plant 50 billion trees in the Middle East and employ new technology to capture and store emissions, aiming for a 'circular carbon economy' (Kennedy, 2021). The country is striving to be a 'green leader' while simultaneously being the world's main global oil producer (Ottaway, 2021). However, the necessary technology is highly expensive, and it is uncertain how quickly it can become sufficiently efficient. On top of that, a study from 2019 found that carbon capture system technology would potentially only reduce emissions by 10-11 per cent (Jacobson, 2019). In august 2023, the United Nations called out Saudi Arabia's 'greenwashing', stating that the company's goals do not align with the Paris Agreement and contribute to spreading misleading information about climate change and climate actions. The United Nations publicly warns Aramco that their plans to increase oil production exacerbate 'climate-fuelled human rights violations' (OHCHR, 2023). Clearly, the excessive optimism in future technological solutions can act as a justifier of current inaction in reducing fossil fuel production, thus posing a severe risk as the planet rapidly warms up.

It is not only fossil fuel producers and companies that take part in mis- and disinformation spreading. State leaders are also directly and indirectly involved in funding disinformation about climate change. For instance, Russia has a long history of denying climate change, with many state-owned news channels promoting climate-sceptical narratives. Russian President Putin has previously made statements claiming that volcanic eruptions produce more emissions than human activities and that global warming could have beneficial effects on a relatively cold country like Russia (Vrba, 2023). Despite this, in 2021, Russia announced a climate strategy to achieve net-zero emissions by 2060. However, no serious efforts have been made by Russia to reduce its emissions. In fact, Russia's energy strategy for 2035 mainly focuses on increasing fossil fuel extraction, consumption and export. In 2022, Russia unveiled a more detailed climate plan to achieve its target, assuming that Russian forests will store more than double the current amount of carbon (Climate Action Tracker, 2023c). Moreover, Russian news channels have alleged that the West is using the climate change debate to hinder Russian economic development (EUvsDisinfo, 2023). Polls on public attitudes towards climate change indicate that the Russian public is much less concerned about climate change compared to the majority of the population in EU countries (Kurbanov and Prokhoda, 2019).

Whether climate misinformation is funded and disseminated by the private or public sector, it influences people's perception of climate change and contributes to slow or minimal action. Ultimately, politicians risk losing support for concrete and urgent climate actions if voters are uncertain about the causes of climate change and the necessary mitigation measures. In the United States and Australia, approximately 33 per cent of the population believes that climate change is natural and not caused by human activities (CAAD and CAN, 2022). In India, around 57 per cent of the population believes that fossil gas is a climate-friendly energy source. In several countries, including Australia, Brazil, Germany, India, the United Kingdom, and the United States, one-fourth of the population believes that their respective country cannot afford to achieve net-zero emissions by 2050 (CAAD and CAN, 2022). Mistrust in science, the spread of false climate information and climate polarisation can proliferate rapidly through social media and algorithmic systems. The rapid spreading of false information can have huge effects on people's perception of climate change, and the actions taken by political leaders. Nevertheless, the message from climate scientists remains unequivocal: greenhouse gas emissions accelerate global warming, and the world must therefore drastically and urgently reduce these emissions.

POWER OF LOBBYING

The influence of lobbying on protectionist, incrementalist and status quo approaches is substantial. In climate strategy, a protectionist approach refers to prioritising national interests and industries, often leading to the implementation of trade barriers and tariffs that can hinder global cooperation and the flow of green technologies, potentially slowing down global climate mitigation efforts. On the other hand, the incrementalist approach involves making gradual, step-by-step changes in environmental policies and actions, focusing on achievable, small-scale improvements over time, rather than implementing sweeping, comprehensive reforms in response to climate change. In the context of climate policymaking, maintaining the status quo means continuing current policies and practices without making significant changes or advancements, often leading to insufficient action in addressing the urgent challenges of climate change.

While countries and political leaders often have failed to prioritise the climate issue, numerous companies have adopted high-profile greenwashing strategies, presenting themselves as environmentally conscious. Alongside the spread of misinformation and its effects on climate actions, lobbying by vested interests presents a significant challenge. In 2019, Forbes reported that the world's five largest publicly owned oil and gas companies spent approximately US$200 million annually on lobbying against binding climate-motivated policies (McCarthy, 2019). They outspent nearly all international non-governmental organisations (NGOs) advocating for a swift transition to green energy. The number of industry lobbyists at international climate negotiations has also significantly increased. Lobbying is a legitimate tool for companies to use during international negotiations, playing a vital role in democratic policymaking by conveying companies' perspectives to decision-makers. However, lobbying must be conducted fairly, justly and inclusively, prioritising the survival of the planet over corporate or client country interests. Large companies have often successfully lobbied to restrict the scope of climate laws and policies. Lobbying can take both indirect and direct forms. The funding of climate-sceptical think tanks represents an indirect form of lobbying. Direct lobbying occurs when lobbyists engage directly with decision-makers. Vested interests have frequently been behind successful lobbying efforts aimed at limiting the scope of climate laws and policies. Reports indicate that during recent

Conference of the Parties (COP) events, there has been a surge in lobbyists from the fossil fuel sector. In COP27, the number of industry lobbyists increased by 25 per cent compared to the previous year (ISD, 2023). They heavily lobby to water down statements about coal and fossil fuel use as measures to mitigate climate change (Hill, 2022). This surge of industry lobbyists hampers urgent action to significantly reduce fossil fuel combustion and associated emissions. In some cases, lobbyists push the planet's climate to dangerous extremes by blocking or diluting policies that would facilitate the transition from fossil fuel-based energy to clean energy sources.

In 2022, US-based fossil fuel companies topped the list of the world's most obstructive organisations in terms of climate policy engagement (InfluenceMap, 2022). These companies have a long history of influencing policymaking through lobbying. For instance, in 2010, the American Clean Energy and Security Act, also known as the Waxman-Markey Bill, was proposed to the US Senate. This act was a significant climate regulation proposal in the United States at the time, designed to promote a clean energy economy, reduce emissions and enhance energy independence (Legal Information Institute, 2021). However, the bill did not pass the Senate in 2010, in part due to extensive lobbying efforts. Research indicates that lobbying by companies reduced the likelihood of the bill's passage by 13 per cent (Meng and Rode, 2019). Another example of corporate lobbying against climate policies is the weakening and restricting of the European Union's Emission Trading System (ETS). This system is a key climate policy within the European Union, aimed at reducing greenhouse gas emissions from industries and energy companies. Energy-intensive industries like steel, cement and chemicals have lobbied heavily to influence the ETS and negotiate exemptions from certain aspects of the regulation. The steel industry, for instance, successfully obtained exemptions from annual updates to the ETS benchmarks (Van den Plas and Martellucci, 2021). Undoubtedly, lobbying can delay urgent climate action by undermining efforts to hold companies accountable for significant emissions reductions.

Conflicting interests, weak political will, lobbyists and the spread of misinformation have resulted in 'minimalist' changes. However, incremental changes that maintain the status quo are insufficient to address the climate crisis. In a world that urgently needs to reduce emissions to combat climate change, there is no justification for continued or increased production of the main contributor to climate change. Delaying tactics also work against the interests of companies, exposing them to greater risks related to weather events and unpredictable energy policies in the future. True and sustainable success will arise from transformational changes in countries and societies, requiring a re-evaluation of their lifestyles and a commitment to global common goals.

The Paris Agreement is a ratified commitment by nearly all countries worldwide. Consequently, companies also have responsibilities to contribute to its effective implementation, as they are part of these countries. Instead of lobbying against effective climate policies, companies should assume moral and social responsibility by advocating for adaptation and mitigation policies that genuinely commit to reducing global warming. Ultimately, companies can contribute to transformational changes by utilising lobbying as a tool to facilitate the private sector's contribution to national and global climate goals for transformational change. There is a growing trend among company coalitions and trade associations that recognise the importance of responsible lobbying, seeking to establish a global standard for ethical climate lobbying (Sullivan et al., 2023). In 2022, chief executives of 154 major global businesses wrote an open letter to the President of the European Commission, urging the European Union to prioritise a green transition for achieving energy security (Corporate Leaders Group, 2022). Though there is some positive lobbying by the private sector, it is still a matter of serious concern that lobbying by the companies and trade associations, particularly from the high-carbon sectors, has become a significant barrier for the countries in undertaking ambitious commitments and agreeing to effective policymaking towards achieving climate goals to which they had agreed in 2015.

To achieve transformational changes, the world must commit to broader environmental objectives. The planet is currently facing an extensive environmental crisis, including biodiversity loss, natural resource depletion, deforestation and water pollution, in addition to climate change. While climate change can exacerbate these environmental issues, it is not the sole cause. Inadequate policies, unequal resource distribution, overexploitation, unsustainable development and rapid population growth are also significant contributors to these environmental challenges. Consequently, even if the world successfully limits global warming to 1.5°C, it will not alone resolve the global environmental crisis. At the same time, the planet is facing a cascade of environmental crises, including alarming biodiversity loss from habitat destruction, pollution, overfishing, deforestation, ocean acidification threatening marine life, critical water scarcity and pollution and soil degradation, all of which demand a concerted global effort and a shift towards sustainable practices and policies. Political leaders often prioritise short-term goals due to their limited tenure, focusing on policies that yield immediate results rather than addressing long-term environmental challenges. Economic interests and lobbying from industries like fossil fuels, agriculture, and fishing, which contribute to environmental problems, exert significant influence on political decisions, often leading to resistance against necessary environmental regulations.

The complexity of environmental issues, which are interconnected and require comprehensive solutions, can result in paralysis or incremental actions that fail to tackle the root causes. This complexity, coupled with a lack of global coordination owing to differing priorities and ideologies among nations, further complicates the formulation of effective global environmental policies. Public perception also plays a role; environmental issues, often seen as less immediate than concerns like the economy or healthcare, receive less attention, reducing pressure on leaders for urgent action. The fear of economic disruption and job losses from implementing environmental policies makes leaders hesitant, while in some cases, a lack of scientific understanding or denial of issues like climate change limits their ability to recognise and address these crises effectively. Additionally, vested interests and corruption in some regions exacerbate environmental degradation, as seen in practices like illegal logging and unregulated fishing. Political leaders bear the complex responsibility to simultaneously address all environmental challenges to make meaningful progress towards creating sustainable societies for all.

SOME PROGRESS

Undoubtedly, progress in international climate negotiations and national commitments has been slow and inadequate. It has struggled to keep up with scientific advancements. While many climate negotiations have been less than encouraging, the world cannot afford to lose hope and abandon its determination to persevere. The Paris Agreement marked a pivotal moment for the world's transition to a low-carbon economy. Despite the bleak outlook based on current and planned commitments and actions, the Agreement has instilled hope that achieving climate neutrality before the end of this century is feasible (Swain, 2020). In the past two decades, several progresses have been made in forging agreements on how to limit and adapt to climate change. These agreements range from bilateral agreements to global ones. The UN-sponsored climate conferences have played an important role in raising global awareness about climate change and promoting sustainable development.

In recent years, there has been an increase in policies and regulations addressing climate mitigation. The IPCC emphasised in its Sixth Assessment Report in 2023 that renewable energy and other mitigation strategies are becoming more accessible, cost-effective and increasingly supported by the public. However, the expansion of renewable energy is predominantly visible in countries with developed economies in the Global North, while it lags behind in several developing economies in the Global South, primarily due to limited financial resources and technological development. Although significant investments are being made in climate mitigation strategies, global efforts need to accelerate to meet climate targets (IPCC, 2023a). In climate negotiations, the world has agreed on collective action, where wealthier industrialised nations bear the responsibility of financially supporting climate actions in less affluent countries. During the 2009 Copenhagen Climate Conference, wealthy industrialised nations pledged to provide US$100 billion in annual climate financing by 2020 (OECD, 2022). The Paris Agreement further introduced specific and coordinated demands for policy changes to enhance climate financing for Global South countries through multilateral financial institutions.

Nevertheless, world leaders have yet to agree on the specifics of climate financing, nor have they fulfilled the promised US$100 billion in funding for climate actions (OECD, 2022). Regrettably, a substantial gap persists between the commitments of rich industrialised countries and their actual contributions to climate action. These nations have been hesitant to engage meaningfully with the demands of low and middle-income countries for debt relief and increased financing for climate action while expediting their own transition to sustainable practices and setting more ambitious climate targets. Given the global climate crisis, rich industrialised countries need to go beyond mere promises and fulfil their existing commitments. During COP27 in Sharm el-Sheikh, new promises were made to establish a Loss-and-Damage fund to finance countries vulnerable to climate change. While the success of this agreement remains uncertain, it suggests that collective progress for climate action is still possible.

COP27 also kindled hope with the resumption of climate discussions between China and the United States. Collaboration between these two major powers is pivotal for a genuine opportunity to limit global warming. Simultaneously, the world appears more divided than ever due to the conflicts in Ukraine and Gaza. In these uncertain times, it is crucial to remember that, despite widespread fear and repeated warnings from policymakers and security analysts, the world has not yet witnessed countries waging wars solely due to climate change. Therefore, hope persists that the world can unite for climate action. Globally and regionally, numerous institutions and agreements have been established to address and mitigate security challenges stemming from climate change (Swain, 2020). The world is undeniably moving slowly, albeit consistently, towards transforming shared concerns about climate change into greater global cooperation. The increasing awareness of climate change and the growing demand from the public for tangible actions against it are starting to compel reluctant political leaders to come together and take some effective, coordinated measures for climate mitigation and adaptation. Civil society movements, with their increasing influence and global networks, play a role in strengthening international policies and programs related to climate change. These collective global actions are crucial to tackling the intricate and widespread challenges posed by climate change. However, successful endeavours to combat climate change must extend beyond minor adjustments and instigate significant transformation in governance structure and policymaking processes worldwide.

SUMMING-UP

The climate crisis, posing a severe threat to our societies, is not being adequately addressed by current political responses due to a combination of factors. Political apathy and self-preservation impede urgent actions needed to prevent extreme outcomes. While leaders convene annually to discuss climate strategies, their commitments fall short of the necessary emission reductions and sustainable practices. The Paris Agreement of 2015, aiming to limit global warming to well below 2°C, marks a significant political response, but the continuous rise in temperatures and emissions highlights the failure of political efforts over the last five decades. Despite minor progress, the urgent need to phase out fossil fuels is largely unaddressed at the international level, leading to a lack of meaningful action at the national level. Furthermore, geopolitical tensions and the absence of key world leaders in climate negotiations have diverted focus from this critical issue. The ongoing rise in greenhouse gas emissions, driven primarily by the continued use of fossil fuels, and the conflicting actions of global leaders underscore the inadequacy of current political responses to the climate crisis.

Protectionism, characterised by trade barriers and national interests, hinders global climate efforts by impeding the international flow of green technologies and slowing the adoption of renewable energy sources. The blame game among countries, assigning responsibility for emissions, further complicates progress. While major emitters like China and the United States are often blamed, considering per capita and historical emissions reveal a more nuanced picture. Political leaders are aware of climate risks, and the majority of the global population is concerned about climate change. However, despite public sentiment, climate issues are often not prioritised in electoral processes, leading to a mismatch between voter preferences and political action. Additionally, climate change has become a partisan issue, with leaders wary of taking measures that may carry electoral risks. While campaigns may promise climate action, the realities of governance often involve compromises that may not align with long-term climate goals. The lack of political will and resistance from industries tied to climate change contribute to the delay in taking decisive steps. The growing youth movements have amplified the urgency of climate action but have yet to overcome these political challenges. Ultimately, addressing climate change requires strong and wise political leadership, global cooperation and a shift in priorities to secure the future.

The spread of misinformation and disinformation about climate change, fuelled by powerful economic and political interests, poses a significant barrier to effective climate action. Misinformation, which includes false or inaccurate information spread unintentionally, and disinformation, which is false information spread deliberately to deceive, have been used by both private and public sectors to sow doubt regarding climate science and delay urgent actions. Major fossil fuel producers, such as ExxonMobil, have been aware of climate change for decades but funded campaigns to discredit it. State-owned oil companies like Aramco have also invested heavily in influencing climate science and public perception. This misinformation and lobbying by vested interests have led to public confusion, complacency and delayed action on climate change, hindering the global response to the crisis.

Despite these challenges, there have been some positive developments, such as the Paris Agreement, which has instilled hope for a transition to a low-carbon economy. Progress has been made in renewable energy adoption, and international climate conferences have raised global awareness. However, there remains a gap between commitments and actual contributions to climate action, especially in terms of climate financing for less affluent countries. Rich industrialised nations need to fulfil their promises and support climate actions in developing countries. Collaborations between major powers like China and the United States offer hope for meaningful global climate action. While challenges persist, there is a growing awareness of climate change, public demand for action and global cooperation efforts that are crucial for addressing this complex issue effectively. Because of all this, it is important that countries consider climate change as a national security threat to be able to attach importance to the issue in order to cooperate with other countries and compromise with their immediate interests.

3
GLOBAL MILITARIES AND THE CLIMATE CRISIS

In this chapter, the focus has been to explore the intricate and often overlooked interplay between military operations and climate change, highlighting both the military's role in environmental degradation and its vulnerability to the changing climate. Historically, the military's presence has left indelible marks on the environment, not only through the well-documented impacts of war and conflict, such as loss of life, displacement of populations and economic upheaval, but also through more insidious means. The military's extensive use of fossil fuels and resultant greenhouse gas emissions contribute significantly to global climate change, a factor frequently overshadowed in public discourse. This chapter also explores the burgeoning debates around the concept of 'greening the military,' analysing the pressures and challenges faced in reducing the military's climate footprint. In addition, it assesses how climate change, in turn, influences military strategies, readiness and expenditure, underlining the bidirectional nature of this critical relationship. As the world grapples with increasing climate instability, understanding and addressing the military's role in this dynamic becomes ever more crucial.

THE MILITARY, ENVIRONMENTAL DESTRUCTION AND CLIMATE CHANGE

Militaries have always affected and been dependent on the surrounding environments. Besides the countless adverse consequences of military activities and armed conflicts (including loss of lives, population displacement and economic disruption) militaries are also responsible for the physical and biological destruction, using large amounts of fuel and contributing to massive greenhouse gas emissions. While various countries and sectors have been named and shamed for their emissions, people rarely point the finger at the military, which is a significant culprit (Klare, 2019).

Military emissions are a global problem as most countries in the world have a military to protect their national security. There are only 31 countries without a military, and these are primarily small island countries. On the other hand, 172 countries have kept active armed forces, and the world spends more than 2.2 trillion dollars on them annually (SIPRI, 2023a). The United States spends the most on their military, and its fuel usage alone contributes more greenhouse gases than the total reported emissions of countries such as Portugal or Sweden (Crawford, 2019). Not only do military equipment, vehicles and weapons emit exhaustive amounts of greenhouse gases, but important carbon sinks that absorb carbon, such as soils and forests, can be degraded and destroyed as a consequence of armed conflicts and military operations. This kind of environmental destruction further increases greenhouse gases in the atmosphere and ocean because it limits the earth's capability to capture and contain them. Altogether, there is no doubt that the world's militaries are significant contributors to global greenhouse gas emissions, directly and indirectly.

Militaries cause environmental destruction at different levels of their operations. Ongoing wars cause massive humanitarian disasters and environmental crises. Besides this, the consequences of previous wars, preparation for wars and threats of hazardous future wars are also ways the environment can be affected by military activities (Swain, 2013). Despite increasing understanding of militaries' environmental footprints in recent decades, armed conflicts continue to inflict severe environmental degradation and destroy the ecosystem. Few attempts have been made to protect the environment from destruction and degradation in armed conflicts. In 2022, after ten years of work and negotiations, the International Law Commission adopted 27 draft principles on the protection of the environment in relation to armed conflicts, recognising the need for urgent action to 'prevent, mitigate and remediate harm to the environment before, during and after an armed conflict' (UN, 2022). The principles include the obligation to remove toxic and other hazardous remnants of war, measures to prevent transboundary harm and promotion of sustainably using natural resources in the occupied territory, among others. Nevertheless, considering the vast environmental destruction caused by armed conflicts and other military activities, the international community still has a long way to go to protect the environment from the destruction and damages caused by previous wars, preparations for wars and ongoing wars.

Previous wars can result in a vast number of unexploded weapons and massive physical and biological destruction of soil and landscapes. For instance, in Lao PDR, which is the most heavily bombed place per capita in the world, farmers are still suffering from the consequences of the US bombings of two million tonnes of explosives in the 1960s and 1970s (Guo, 2020). Around one-third of the bombs did not explode, still contaminating around 25 per cent of all villages and 50 per cent of all agricultural land in the country (World Bank, 2007). During the same war, large areas of forests were destroyed. According to Arthur Westing, one of the pioneers in this field, nearly two million hectares of Vietnamese forest and 300 thousand hectares of cropland were destroyed due to the bombing and spraying of the herbicide 'Agent Orange' (Westing, 1972). This strategy was used to eliminate hiding places in the forest and destroy food production for the adversaries. Populations in the war-affected areas still suffer from health consequences from contamination caused by the war. Based on Population Census data from 2009, ethnic minority women, who had a marginal possibility to migrate from contaminated areas, report a higher rate of disabilities than women in less contaminated areas (Yamashita and Trinh, 2022). Undoubtedly, even wars that ended several decades ago can still have severe effects on the human populations and ecosystems in the affected areas.

In addition, previous wars have accounted for huge amounts of greenhouse gas emissions that still are trapped in the atmosphere as most emissions stay there for around 300–1,000 years (NASA, 2019b). A recent estimation by Professor Neta C. Crawford shows that only in 1975, in the last year of the Vietnam War, the US military emitted at least 109 $MtCO_2e$ (million tonnes of carbon dioxide equivalents), which is equivalent to providing energy to 13.12 million homes for a year (Crawford, 2022). This estimation does not even include emissions released due to the large extent of forest loss. Moreover, the post-war reconstruction of destroyed infrastructures also contributes to increased emissions. Estimations show that reconstruction after wars with large-scale destruction on a national level can lead to more than 100 million tonnes of CO_2 emissions (Michaelowa et al., 2022). Therefore, even though peace may be achieved between nations, the aftermath of previous wars can still have devastating effects on the environment and human populations for a long period after the war has ended.

Military activities *preparing for war* can also contribute to environmental degradation through large expanses of fragile land, marine resources and air space being bombed or polluted. Globally, military training areas cover five to six per cent of the total land surface (Zentelis and Lindenmayer, 2015). The production, testing and storage of conventional, chemical, biological and nuclear weapons create toxic and radioactive substances, contaminating soil, air and water. Between the 1970s and 2003, the US Navy used parts of Vieques Island in Puerto Rico as an explosive test site. It was used as a target practice for the naval warships using live ordinances that included napalm and even depleted uranium for over three decades. Depleted uranium is the primary by-product of uranium enrichment and is a mildly radioactive metal used in munitions for its penetrating ability. Despite opposition from environmentalists, the fisher community and the local population, the naval exercise continued to contaminate the island's air, water and soil. Two decades have passed since the naval exercise was halted, but substantial work remains to clear up the land surfaces suspected of having munitions. With a total population of around 10,000, the island has the highest sickness rates in the Caribbean (Pelet, 2016). Inevitably, even countries that are not in war risk being heavily affected by the dangers of contamination from munitions if sites in their territory are used for military training purposes.

Among all the military preparedness acts, the adverse effects of nuclear weapon production and testing are the most severe and enduring. The difference between testing conventional and nuclear weapons is that the vestige of the nuclear tests lasts forever. Since the end days of the Second World War, the United States, the former Soviet Union, the United Kingdom, France and China tested their nuclear weapons in the open deserts and the sea until the partial test ban treaty in 1963 which moved the nuclear tests underground to restrict the spread of radioactive clouds to some extent. The nuclear tests that the United States had conducted in some of the Pacific islands during the Cold War still suffer from high radiation levels and remain uninhabitable. The United States conducted most of its nuclear tests, almost 1,000 in the Nevada desert. The Soviet Union conducted 456 nuclear tests in Kazakhstan's Semipalatinsk. The population in that region still suffers from high cancer rates as a consequence of radiation exposure. In 2011, the director of the Semey Radiological Institute reported that the cancer rates in the region exceeded the national average by around 200 per cent (CTBTO, 2011). Moreover, long-term environmental consequences of nuclear tests are radioactive contamination of surface soil and groundwater, making previous test areas unsuitable for habitat or agriculture. The dangerous legacy of nuclear testing continues to affect many poor and underprivileged communities in all countries with nuclear weapons. This makes both the preparations for and the potential future nuclear, biological or chemical warfare a serious threat to human lives and the environment.

Ongoing armed conflicts are significant contributors to environmental destruction and substantial amounts of greenhouse gas emissions. During ongoing armed conflicts, essential water, energy, sanitation and waste infrastructure can get damaged. In some cases, environmental damages get carried out as the deliberate objective by warring parties to increase the enemy's vulnerability, rather than an unwanted by-product. For instance, in armed conflicts in the Middle East and North Africa region, attacking and taking control of essential environmental infrastructures are becoming increasingly common warfare by state and non-state militaries (Sowers et al., 2017; Weinthal and Sowers, 2023). In Syria and Iraq, Islamic State (IS) has used water and energy installations as a means of warfare (Swain and Jägerskog, 2016). To some villages, water and energy were cut off. To others, water was deliberately contaminated, polluted or controlled to flood the adversaries. In 2014, IS took control of a dam in Falluja, Iraq. They released an extensive amount of water which submerged governmental facilities and flooded 200 km^2 of fertile farmland, destroying almost the entire agricultural yield, killing livestock and causing forced displacement (Swain and Jägerskog, 2016). Between 2011 and 2021, water infrastructure was targeted at least 180 times in Gaza, Libya, Syria and Yemen (Borgomeo et al., 2021). Additionally, Israeli military operations in Gaza have been targeting water infrastructure desalination plants and water delivery systems quite often. In the war between Israel and Hamas starting in October 2023, the Israeli military forces have been accused of completely blocking the supply of water entering Gaza. These kinds of attacks on the water supply system can be considered as weaponising the water. Cutting off access to water, polluting water and generating flooding as warfare risks having deadly outcomes, and is a particular risk for the most vulnerable parts of the population.

Armed conflicts are also accountable for massive deforestation and the destruction of wildlife, directly or indirectly, such as the large forest cover destroyed in the Vietnam War. In 2020, the world witnessed a ten per cent increase in deforestation in conflict-affected areas. Armed conflicts can trigger forest loss in various ways. Similar to the destruction of essential infrastructure, the destruction of forests targeted in air strikes can be a deliberate objective to increase the vulnerability of the enemy. In Syria, forest cover has decreased rapidly due to the ongoing civil war, exceeding the global average forest loss rates between 2000 and 2019. One of the main drivers of deforestation is forest fires triggered by bombings from different parties in the conflict (Mohamed, 2020). In addition, in 2018, armed opposition groups in northwestern Syria cut down more than half a million olive trees owned by Kurdish parties as a means to weaken the adversary's possibility to sustain themselves while increasing the opposition group's revenue by selling the collected wood (Shkaki, 2020). In conflicts, forests can become a highly valuable resource for state and non-state militaries as well as for civilians. Wood and timber can be a critical income to build up the supply of military ordnance. In Myanmar, the military rule has cut down large areas of highly valuable teak trees to export to bordering countries to increase their revenue and thereby also strengthen their power position. Similarly, in the Democratic Republic of Congo (DRC), illegal logging in the rainforest is carried out by armed groups for charcoal production and illegal mining purposes. Moreover, the conflictual situation in the country forces civilians to migrate deep into the forest for protection. The forcibly displaced population becomes dependent on farming, producing charcoal and using wood as fuel to sustain themselves. Therefore, deforestation also occurs deep into the forest as armed conflicts push populations to new areas in search of survival (Darbyshire, 2021). The tropical rainforest in the DRC is highly important to regulate the climate in the Central Africa region. Deforestation can substantially alter the water cycle and thus affect local and regional precipitation patterns. It is estimated that continued deforestation in the DRC will reduce local precipitation by eight to ten per cent by 2100 (Smith et al., 2023). Tropical forests are of major importance to the carbon cycle as they absorb and store extensive amounts of CO_2. Thus, deforestation in these areas is particularly devastating for the climate. In 2020, deforestation in tropical areas experiencing armed conflicts caused 1.1 mega tonnes of CO_2 emissions, which is equivalent to almost four times the total reported emissions in the United Kingdom during the same year (Darbyshire, 2021).

Moreover, military activities such as the circulation of heavy military vehicles and explosions increase toxic elements in the atmosphere and the soil (Pereira et al., 2022). In 2021, the US forces left Afghanistan in a hurry but left behind a terrible toxic environmental legacy. The United States and its allies had been using many military facilities in Afghanistan for nearly two decades. These deserted facilities have left behind several burn pits and dangerous chemicals destroying soil and polluting water that can produce long-lasting health and environmental hazards (Atherton, 2021). The contamination from the US military on Afghan soil and its consequences on the human population and ecosystems have never been fully investigated or addressed. However, the severe consequences risk lasting for several generations as chemical, biological and medical waste never have been cleaned up (Billing, 2023).

Environmental hazards such as pollution of radioactivity and chemical toxicity can dramatically affect soil's properties and destroy soil for centuries (Pereira et al., 2022). Soils, just like forests, are majorly important for the climate because they absorb CO_2 emissions that otherwise would be emitted into the atmosphere. Soil's carbon storage capacity is even larger than the forests, and three times larger than the atmosphere. Degraded soil reduces its capacity to absorb carbon, thus increasing the emissions released into the atmosphere which intensifies global warming (Onti and Schulte, 2012). On top of that, soil degradation also increases the vulnerability of ecological systems and thereby reduces the available options to adapt to the changing climate, especially agricultural production. For instance, in the Russia-Ukraine war, soil degradation due to chemical toxicity and radioactivity from the armed conflict can lead to adverse global consequences. Though the war's full effects on soil degradation are still unknown, it can create catastrophic food insecurity worldwide as Ukraine has one of the most fertile soils globally and is one the top grain exporters in the world (Pereira et al., 2022; Rawtani et al., 2022). Consequently, ensuring food security and ending all forms of hunger by 2030 becomes a distant goal to achieve.

Armed conflicts are also responsible for detrimental air pollution. In the Russia-Ukraine war, massive amounts of toxic gases are emitted into the atmosphere from the bombing of ammunition depots, fuel depots, chemical industries and refineries (Pereira et al., 2022). In Ukraine, many of the bombed sites are located close to residential areas. Except for the immediate threat to life, the bombings cause severe air pollution which can cause serious health issues such as cancer and reproductive issues. Considering that air pollution is the main environmental health risk globally and that the air quality in Ukraine already was among the worst in Europe before the Russian invasion in 2020, this has harmful effects on soldiers, civilians and ecosystems (UNDP, 2022). While the direct effects of military toxins on the environment are limited and localised in nature, the military operations near nuclear power stations can cause a much more widespread and severe problem of radioactive pollution crossing Ukraine's borders. In addition, the aggravated air pollution caused by the war increases greenhouse gas emissions into the atmosphere, directly contributing to intensified climate change.

GREENING THE MILITARY

There is no doubt that the military is responsible for a huge amount of greenhouse gas emissions. Currently, the lack of obligation and transparency in reporting the military's greenhouse gas emissions makes it difficult to estimate all the military emissions worldwide. Based on estimations of available information, military bases, equipment, transportation and arms contribute at least five per cent of global greenhouse gas emissions. The five per cent does not even include the massive release of greenhouse gases from wars. If those emissions are included, the share goes up to at least six per cent (Parkinson, 2020). This means that if all the militaries combined were a country, it would have the fourth largest national climate footprint worldwide, after China, the United States and India (Parkinson and Cottrell, 2022). The emissions from the military sector are so huge that it is even higher than that of civil aviation (Parkinson, 2020). Despite its large contribution to global greenhouse gas emissions, reducing military emissions is often left out of global climate change negotiations and discussions. Nevertheless, the military sector has played a major role in funding climate science to reach the current understanding of climate change and its impact on our societies.

During the Cold War, NATO, and in particular the US military, funded scientific work specifically focusing on meteorology, oceanography and radio meteorology to improve their military operations (Turchetti, 2018). One of the key findings from the funded scientific work was the linkage between accumulated CO_2 in the atmosphere and ocean and global warming. Climate change was recognised as a threat to our societies, having the potential to create wars and societal unrest. Despite being at the forefront of obtaining knowledge on climate change and global warming, militaries have shied away from taking measures to mitigate it. In fact, militaries have been a strong driving force in developing techniques dependent on fossil fuels during the 1900s, to strengthen military capacities (Crawford, 2022). The NATO Scientific Committee has also pushed for technological solutions to environmental problems, including issues related to the changing climate (Turchetti, 2018). However, reducing emissions and engaging in the green energy transition have only recently started to be acknowledged as explicit goals within the military sector itself. Instead, the main focus has been on adapting military bases and operations to climate change. While global climate change poses an existential threat to the planet, nation-states have remained preoccupied with protecting and promoting their national interest via their military.

When the Kyoto Protocol was adopted in 1997, world leaders committed to reporting their country's emissions to strive for emission reduction to prevent the irreversible effects of climate change. During the negotiations, pressure from the US led to the exemption of the military from emission targets. The Department of Defense warned that including military emissions in the report would lead to emission cuts that would jeopardise their national security. At the same time, the military establishment of the West assumed that climate change would pose unconventional security challenges to them by creating chaos, societal violence and mass migration in poor and developing countries. With this narrative, the threat that climate change poses to national security even increases the relevance of investing in the military in order to better protect the nation from future threats posed by climate change (Crawford, 2022; Turchetti, 2018). In other words, the threats of the effects of climate change can be used as an argument to strengthen military capacities, while the military's own responsibility to reduce emissions is nominal.

When the Paris Agreement was adopted, it was decided that the individual member states determine if they would include the reduction of military emissions in their total emission reduction reports. In other words, reporting military emissions is voluntary. The countries choosing to report their military emissions often report insufficient or unclear numbers, as it may only include certain parts of the emissions from the military, or be presented together with emissions from other sources (Parkinson and Cottrell, 2022). For instance, the United Kingdom is one of the best countries worldwide at tracking their military emissions, but their reporting is only disaggregated to the point where emissions from aircraft and naval vessels are clearly demonstrated, whereas emissions from bases and ground vehicles are not distinctively reported (Rajaeifar et al., 2022). Yet, the United Kingdom and the other NATO allies are members of the Organization for Economic Co-operation and Development (OECD). They regularly publish data on energy consumption and economic activities, making it possible to estimate their military greenhouse gas emissions somewhat reliably (Parkinson, 2020). Similar data from other countries with large militaries like China, Russia, India and Saudi Arabia are not easily accessible.

However, if any meaningful reduction in military emissions is to be achieved worldwide or within the NATO bloc, the place is to start with the United States, which spends most on its military globally. The US military is the single largest institutional producer of greenhouse gas emissions of any sector worldwide. Though US military emissions only account for one per cent of the country's total amount of emissions, the military is still accountable for almost 80 per cent of all the emissions from governmental institutions in the country. In fact, no other single agency uses more petroleum than the US military, with its 500 domestic and 75 bases worldwide (Crawford, 2022). For the last two decades, the US military has been discussing its concerns over climate change. In 2003, when the Bush administration was in climate denial mode, the Department of Defense even brought out a report stating how global warming can incite food, water and energy insecurities, in turn, having the potential to lead to geopolitical tensions or armed conflicts. In effect, the Department of Defense argued that climate change should be elevated to a national security concern (Schwartz and Randall, 2003). The Obama administration indirectly worked to reduce carbon emissions of the US military by trying to cut down the fuel costs for military operations in Iraq and Afghanistan and also to decrease the number of fuel-transport convoys as those were highly vulnerable to enemy attacks. Between 2003 and 2007, around 3,000 US soldiers were killed or wounded during attacks on fuel and water convoys in Afghanistan and Iraq. In Iraq, one in every eight soldier fatalities was connected to fuel transit or protection during that period. In 2010, ground convoys were attacked at least 1,100 times in Afghanistan (DoD, 2011). In addition, the spending on fuel had increased by over 200 per cent compared to the decade prior. Around 15 billion dollars were spent on energy, which of 80 per cent was spent on oil (Daniel, 2011). In other words, reducing costs and building stronger operational resilience indirectly became a strategy to reduce emissions by transitioning to renewable energy sources, mainly solar-powered electricity (DoD, 2011). Later, the Trump administration ignored the climate change part in handling the affairs of the Pentagon. However, under President Biden, the Department of Defense created a Climate Adaptation Plan in September 2021 and has elevated climate change to a critical national security issue (DoD, 2021a).

To strengthen its commitment to climate change issues, the US military recently started to track its military emissions for the past ten years and has produced a plan to reduce its emissions. The reports include emissions from their vehicles, installations and energy purchase, but do not include all the emissions the military sector is responsible for. Military-industrial emissions such as from the production of military equipment are not included, even though the United States is the largest producer of weapons worldwide. In 2020 alone, the United States sold more than 175 billion dollars worth of arms (DoS PM, 2021). The production of arms, their use, and their disposal contribute to climate change in different stages of the supply chain, and global arms production and sales are increasing every year. Nevertheless, the climate footprints of procurement of military hardware and other supply chains are rarely considered. For instance, in 2019, the US military reported that they emitted 54.8 $MtCO_2e$. However, this amount does not include emissions related to industries producing equipment and weapons for the military. There are many industries connected to the military-industry complex in the United States. Crawford (2022) calculates that if the twelve largest military industries producing weapons in the United States were included in the report, the total emissions for 2019 would be 105.75 $MtCO_2e$. This amount is equal to the combined emissions of the forty-five smallest emitting countries or the emissions from Sweden, Norway and Denmark combined. Crawford emphasises that even this amount is not complete, as she has not included all the emissions connected to the military industries and activities. Mainly, the reported emissions only account for emissions from energy purchases and fuel usage. In other words, except for military-industrial emissions, all other emissions related to the military sector are excluded. These are emissions related to the transportation and distribution of goods, business travels, reconstructions and investments as well as the release of greenhouse gases from the destruction of forests, soils and land. Nevertheless, US military emissions have reduced in recent years, mainly because of the retreat of troops from missions in the Middle East (Crawford, 2022). Despite this, the US military is still accountable for more greenhouse gas emissions than any other military worldwide.

Given the possibility of estimating military emissions from OECD countries, Parkinson and Cottrell (2021) have calculated the carbon footprint of the militaries in the European Union, as shown in Table 3.1. The calculations are based on military expenditure on production and consumption. Though estimations have been difficult to make due to limited available data, it shows that all EU militaries combined emitted at least 24.8 $MtCO_2e$ in 2019. The amount is equivalent to emissions from 14 million cars during one year. When scrutinising military emissions from six EU countries (France, Germany, Italy, the Netherlands, Poland and Spain) more closely, they found that all countries emit more than what they are reporting. For instance, Germany has reported that they emit 0.75 $MtCO_2e$, while the calculation estimates at least 4.53 $MtCO_2e$. Similar to the case of the US military, it is mainly emissions related to the military industry that is excluded from the report. Direct emissions from stationary and

Table 3.1 Military Emissions in EU Countries

EU Country	Reported Military Emissions ($MtCO_2e$)	Estimated Military Carbon Footprint ($MtCO_2e$)
France	Not reported	8.38
Germany	0.75	4.53
Italy	0.34	2.13
The Netherlands	0.15	1.25
Poland	Not reported	Insufficient Data
Spain	0.45	2.79
EU total	**4.52**	**24.83**

Note: The table shows the discrepancy between reported and estimated military emissions in the European Union and specific detail is provided for the six EU countries investigated by Parkinson and Cottrell (2021).

mobile armed forces are calculated to be 1.5 MtCO$_2$e, and indirect emissions from the military technology industry are 3.03 MtCO$_2$e. However, many of the top military technology companies in Germany do not report their emission, which may have lowered the total greenhouse gas emission estimation. The EU country emitting the most is France, accounting for one-third of all the EU military emissions. Though France has one of the world's largest militaries with troops deployed in Africa, the Middle East and the Asia-Pacific, they have not reported their military emissions explicitly. The estimated climate footprint for France is 8.3 MtCO$_2$e, with most of the emissions coming from the military technology industry. The estimations are based on military expenditure data for the fiscal year 2019 (Parkinson and Cottrell, 2021). The recent years' increase in military expenditure in EU countries has most probably led to an increase in greenhouse gas emissions.

The estimations of the emissions by the US and EU militaries show that they are far away from making a serious impact on mitigating climate change. Despite this, the overarching goal of NATO is to achieve net-zero emissions by 2050 and a reduction of emissions of at least 40 per cent by 2030 (NATO, 2022a). NATO member states have also 'committed themselves to engaging in the energy transition by significantly reducing greenhouse gas emissions from military activities and installations without impairing personnel safety, operational effectiveness, and the deterrence and defence posture' (NATO, 2023). This declaration is laudable, but NATO has neither clarified how it would achieve this ambitious target, nor which emissions that should be included in the reduction targets. Individual NATO countries have set their own targets to reduce their climate footprint. For instance, France's Ministry of the Armed Forces (2022) has set the target to reduce its climate footprint from property occupied by the Ministry by 40 per cent by 2030. By the same year, the Netherlands Ministry of Defense (2021) aims to achieve that at least 50 per cent of the energy used at camps is sustainably generated energy. These targets mainly focus on energy transition which is driven by the desire to improve energy self-sufficiency and military efficiency. Additionally, the decreased dependency on fossil fuels can increase military readiness and mobility because troops will be less vulnerable to fuel shortages and can put less effort into transporting tanks of fossil fuels to distant camps (Crawford, 2022). It can also be a strategy to save money and make mission conditions safer for soldiers as was the case for the US troops in Afghanistan and Iraq during the Obama administration. Thus, 'greening' the military may have a larger agenda than only doing all it can to reduce military emissions for the sake of mitigating climate change.

Moreover, NATO explicitly states that the Russian invasion of Ukraine incites a faster energy transition to more reliable and local energy production which also would decrease the dependency on Russia (NATO, 2022b). However, an extensive energy transition to renewable energy sources such as solar and wind energy requires vast extraction of minerals and metals needed in the new technologies. When fleets are changing to electric vehicles charged by renewable energy, minerals and metals such as cobalt, lithium and other rare earth elements are required in large amounts. The extraction of these minerals and metals is an energy-intensive process and can only be found in a few places. Therefore, the energy transition risks creating new dependencies (Rajaeifar et al., 2022). NATO recognises this issue by stating that the energy transition 'should take into account the importance of not creating new strategic dependencies, in particular on China, which currently dominates the control and processing of essential materials' (NATO, 2022b:3). The desire to be more energy self-sufficient and decreasing dependencies, thus vulnerabilities, can be challenging for NATO because rare earth minerals are not extracted in Europe and are only limitedly extracted in the United States. However, in 2023, Europe's largest deposit with one million tonnes of rare earth metals was found in northern Sweden, which if extracted would decrease Europe's dependency on China and other countries extracting rare earth elements. Nevertheless, it is estimated that it would take at least 10-15 years to begin extracting and delivering raw materials if the extraction is approved (LKAB, 2023). In addition, increased extraction of minerals and metals can cause or exacerbate conflicts and tensions. For instance, the increasing demand for cobalt for the energy transition can be linked to intensifying conflicts and human rights violations in the Democratic Republic of Congo, a main supplier of the rare mineral (Prause, 2020). Also in Sweden, tensions between local indigenous populations, who risk getting their livelihood destroyed by the planned exploitation, and the mining company can be exacerbated if the extraction of rare minerals gets approved. In other words, a full-scale green energy transition is complex and complicated to achieve and can have unintended consequences.

IMPACTS OF CLIMATE CHANGE ON THE MILITARY

In a striking paradox, the military, often seen as a major contributor to greenhouse gas emissions, finds itself in the crosshairs of climate change's relentless march. The very forces they've unwittingly helped unleash are now turning against them, posing unprecedented challenges to their installations and operations. As climate change exacerbates geopolitical tensions, it acts like a catalyst, stirring the pot of global unrest and potentially sparking armed conflicts. This isn't just about the military being called upon more frequently for international peacekeeping missions in regions teetering on the brink of state collapse or grappling with the complex issues of climate-induced migration. There is a deeper, more immediate concern. The military is waking up to the stark reality that climate change is a direct threat to their home turf – their bases, their cutting-edge equipment and their state of preparedness are all under siege by an invisible yet relentless enemy. Rising sea-levels, extreme weather events and shifting geopolitical landscapes are not just scenarios in a strategy game; they are real, palpable threats that are reshaping the military's role and response in a rapidly changing world.

With decades of knowledge about climate change, NATO and the United States have had years to plan for new and unconventional military challenges induced by climate change with more frequent and intense weather extremes. Recurrent flooding, sea-level rise, drought, desertification, heatwaves, wildfires and thawing permafrost threaten military bases, installations and operations in various ways. In some cases, the effects can be so severe that bases may have to be abandoned, installations are destroyed or inundated and the viability and durability of arms and hardware can be significantly reduced.

The naval forces, in particular, are vulnerable to storm surges and sea-level rise. More than 1,700 military installations managed by the US Department of Defense worldwide are at risk of being affected by sea-level rise (CRS, 2019). Critical infrastructure can be inundated, thereby also limiting the use of naval bases temporarily and permanently (NATO, 2022b). For instance, the Naval Station Norfolk area in Southern Virginia is the world's largest naval facility and is at risk of being adversely affected by rising sea-levels and storm surges. Already now, flooding is happening more frequently which disrupts the navy's training and readiness, delays the repair and maintenance of ships and submarines and postpones scheduled operations. By 2050, the sea is projected to rise almost one metre compared to pre-industrial levels in some parts of the world, which will affect the ability to conduct operations in the Atlantic (DoD, 2021b). The NAS Key West in Florida is another key military base for training and operational support that is highly vulnerable to sea-level rise. The military base is low-lying and is built upon porous limestone bedrock, which hampers the ability to build sea walls that would hold back the water. Here, the sea is expected to rise more than one metre compared to pre-industrial levels, which would inundate a majority of the military base (Union of Concerned Scientists, 2016). Though the military is taking measures to safeguard its facilities, the threat of sea-level rise is forcing the military to change its contingency plans to improve its adaptation strategies to climate change (DoD, 2021b).

Except for rising sea-levels, the melting ice in the Arctic region can affect operations because the decrease in ice cover can make submarines more detectable from the air, while it may also reduce the effectiveness of anti-submarine warfare techniques adapted for that region (DN, 2014). Operations in previous ice-free areas can be affected by new challenges to ensure safe operations when ice floats are more fragmented than previously (IISS, 2022). A similar situation is likely to happen in Antarctica, where the sea ice is melting quickly. In 2022, the sea ice in Antarctica had the lowest levels ever recorded (WMO, 2023a). Land operations can also be affected by melting ice and changes in the water cycle. Thawing permafrost can damage and destroy roads, buildings and other kinds of infrastructure because of sinking soil and erosion (Pinson et al., 2020). Heavy precipitation and flooding can impact land operations because transport networks can be disrupted which can impede operations that require various forms of supply and mobility of personnel (NATO, 2022b). This is a serious challenge because by only looking at the 79 main installations managed by the US military, two-thirds are highly vulnerable to future recurrent flooding. One-half of the installations are vulnerable to droughts and another half are vulnerable to wildfires, mudslides or erosion (DoD, 2019).

Drought, wildfires and desertification can have huge impacts on installations as buildings and roadways can be damaged or destroyed. Drought and desertification can also disrupt the water supply to the installations, affecting the possibility of carrying out operations and can have serious impacts on the health of soldiers and civilians (NATO, 2022b). Moreover, decreased soil moisture during periods of drought can create deep cracks in the soil, damaging road surfaces which can impact military readiness (DoD, 2019). Hotter temperatures can decrease the durability and viability of tools, techniques and the performance of the soldiers. To deal with the challenges, increased energy consumption may be necessary in order to maintain the efficient cooling capabilities of facilities and installations to prevent overheating (NATO, 2022b). The need for more maintenance and repair due to damages from extreme weather events can also increase energy consumption. Thus, extreme weather events can even increase military energy demand whether their energy is renewable or not. At the same time, the energy supply system can be disrupted by inundations, storms or wildfires. During 2020, military installations in the United States experienced 3018 unplanned utility outages, of which 25 per cent were due to the force of nature (DoD, 2021c). Intensified and more frequent weather extremes such as storm surges or flooding inundating facilities, risk making unplanned utility outages even more common. In turn, it will require more time and effort for maintenance and repair.

Air forces are vulnerable to changes in wind patterns and dust storms. Dust storms are expected to increase in frequency and intensity when areas become more arid due to desertification (IPCC, 2019a). Dust storms can damage aircraft turbines and engines, decrease visibility and disrupt flight trajectories (NATO, 2022b). Hence, it can impede important missions or be another reason for the more frequent need for maintenance and repair of crucial equipment. Hotter temperatures can also affect air force operations because the air temperature together with air pressure, precipitation variability and wind patterns all impact the capabilities of air force operations. Increased air temperatures can cause more frequent events of overheating and require longer runways for takeoff speed (Crawford, 2022). Changes in wind patterns and precipitation can make air forces experience more frequent and longer waiting periods during air periods and changes in their payload capabilities (NATO, 2022b).

The changing climate not only creates more difficult conditions for military missions and adversely affects the viability and durability of military hardware, but it also exposes troops to new diseases and other health challenges. The physical health of military personnel can be put at risk during periods of extreme heat and drought conditions, especially due to the heavy clothing and packs typically used by the military (Pinson et al., 2020). Hotter temperatures have already caused increased rates of heat stroke and heat exhaustion among military personnel. During the years 2014–2018, more than 11,000 active-duty service members in the US military were affected by heat-related illnesses and death. Most cases were reported in areas where high temperatures coincide with high levels of humidity (Hasemyer, 2019). To prevent heat-related illnesses, more 'black flag days' are expected, when military training is suspended because of the hot temperature. More days with high risks of fire hazards are also expected, which is another reason military training may be suspended. The hotter temperatures also increase the risk of catching vector-borne diseases such as malaria or dengue fever. Warmer temperatures during winter and spring periods reduce the vector mortality rates and prolong the reproductive seasons (Pinson et al., 2020). Climate change is also expected to increase the geographic spread of vector-borne diseases. Currently, vector-borne diseases cause more than 700,000 deaths annually (WHO, 2020). Vector-borne, food-borne and water-borne diseases are all expected to increase as a consequence of climate change and will have adverse impacts on human health on a global level (IPCC, 2022a). Though military readiness in certain areas will be impacted by this, it can have serious effects on all parts of the population in affected areas. Nevertheless, the increasing health challenges require the military to more focus on establishing sufficient and efficient medical supplies and facilities.

The increasing risks of diseases, unusual heat and cold waves, devastating floods, hurricanes and wildfires as well as extended periods of drought and famine are all causes which increase the requirement of military readiness for rescue and relief operations. Militaries have historically played a major role in disaster relief and humanitarian aid. Thus, the increase in the number of natural disasters also increases the number of military missions in response. Worldwide, militaries have been increasingly occupied in national and international rescue and relief operations from flooding, hurricanes and fighting wildfires. For instance, escalations of wildfires have caused the military in Australia, Europe and the United States to spend more time on firefighting tasks. In the United States, the number of personnel days of firefighting increased from 18,000 in 2019 to 172,000 in 2021 (Birnbaum, 2022). In turn, other missions and operations will be impacted by the reduced availability of troop members for their regular security duties. At the same time, changing geopolitical conditions can trigger new armed conflicts which strain troops' availability to help in disaster and relief operations (NATO, 2022b). In 2022, when European militaries were concerned about providing aid for Ukraine, while also putting increased efforts to strengthen their own military power, almost 8,000 km² of forest and land was burning within European territory. Military troops were strained and had to prioritise which tasks to put the most effort into (Birnbaum, 2022). These challenges are expected to increase in the future.

Overall, climate change creates challenges that disrupt access and limit military mobility, readiness and effectiveness. Given all of these challenges, there is an increased demand for the military in different parts of the world to build or renovate their buildings and bases to make them climate-proof and adapt their operations to the changing climate and adverse weather events. However, adapting the military to these challenges is expansive, technically complex and can be time-consuming. Thus, not all militaries have enough capabilities to readily adapt to the changing climate, making some nations more vulnerable towards the increasing threats than others. Additionally, even increased investment in the military to adapt to the risks of climate change can have counterproductive effects because it also risks increasing greenhouse gas emissions if their climate mitigation strategies are not sufficient, and thereby also risks increasing the new threats they want to avoid.

INCREASING MILITARY EXPENDITURE AND CLIMATE CHANGE

While NATO allies and a few other militaries in the Global North are starting to find ways to adapt to the changing climate and now recently set targets to reduce emissions, climate change as a security threat has received limited attention in the Global South. Many low-lying Small Island Developing States have raised concerns about their national security when the sea-level is rising. Therefore, countries such as the Maldives have engaged in 'greening' their defence to reduce their climate footprint (Rasheed, 2022). Growing attention has also been given to reducing military emissions in India, one of the world's top spenders on the military. Until recently, specialised vehicles used by the Indian Armed Forces were exempted from the vehicular emission norms set by the country in an attempt to reduce emissions. In 2022, the Indian Armed Forces adopted a plan to use electrical vehicles within the army (Dongare, 2022; The Economic Times, 2019). In other places, climate change as a security threat has mostly been addressed through the increasing demand for militaries to respond to disasters, such as in Latin American countries (O'Toole, 2017). For instance, in Chile, the Ministry of Defense describes climate change as an increasing security threat where the national defence is responsible for supporting in emergencies and disasters as well as searching for ways to adapt to the changing climate (Ministerio de Defensa Nacional, 2017). Notably, some climate mitigation and adaptation strategies are undertaken by a few militaries in the Global South, though the present emission reduction targets and adaptation strategies are not as ambitious as the ones taken by many militaries in the Global North. Except for NATO's ambition to reach net-zero by 2050, countries such as Australia and Japan have set targets to reduce their military emissions. Australia has committed to achieving net-zero by 2050 as a plan to increase defence capability by reducing dependency on fossil fuels (Cole, 2022). Japan has committed to achieving carbon neutrality by the same year but has exempted emissions from military industries producing defence equipment in the target (Dominguez, 2023).

The top world military spenders like China, Russia, India and Saudi Arabia don't have any explicit target for reducing military emissions, except for the recent efforts to include electric vehicles in the Indian Armed Forces, and some attempts by the Chinese military to save energy. Particularly, China and Russia have increased their military expenditure in recent years by producing more arms and increasing their defence capacities. Together with the United States, these three countries are the top countries spending the most on their militaries, accounting for 56 per cent of the total global expenditure (see Figure 3.1). However, the United States have the highest military expenditure, spending 40 per cent of the global military expenditure. Combined with the other NATO member countries, their military expenditure reaches 55 per cent (SIPRI, 2023a). Worldwide, military expenditure has increased by 3.7 per cent during 2022. Europe experienced its sharpest rise in military expenditure in at least 30 years, an average of 13 per cent increase. The sharpest rise was experienced in Finland, with a 36 per cent increase in military expenditure, followed by Lithuania (27 per cent), Sweden (12 per cent) and Poland (11 per cent). The increasing military expenditure globally is mainly a response to the declining security environment with the heightened threat from Russia, which is portrayed by the sharp increase in military expenditure by countries close to the Russian border in Europe. Increased tension in East Asia is also a reason for the rise in military expenditure, particularly by China and Japan (SIPRI, 2023a).

Increasing military spending is a major risk for increasing greenhouse gas emissions (Parkinson and Cottrell, 2021). Though some efforts are made to transition to renewable and more efficient energy usage within NATO allies, the heightened focus on expanding platforms and training to ensure military effectiveness increases both the demand and consumption of energy and military equipment. Hence, also increases military greenhouse gas emissions (NATO, 2022b). Thus, the 'military greening' commitment conjoins with a significant increase in the European countries' military budgets as a consequence of the war in Ukraine. The rush to strengthen the military firepower diminishes whatever pretension they had to reduce the greenhouse gas emission of their armed forces. The lack of transparency about the method NATO will adopt to go green by achieving its ambitious goal of reaching net-zero emissions by 2050 without impairing operational effectiveness and defence capabilities raises doubts about whether NATO is only engaged in greenwashing,

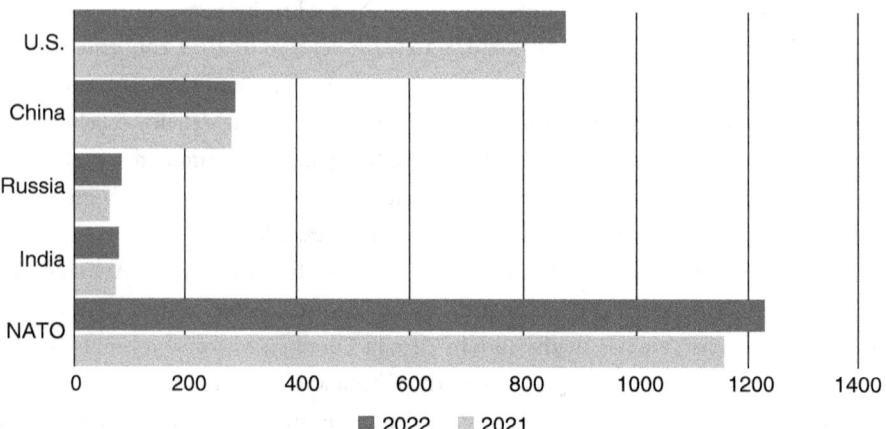

Figure 3.1 Increasing Military Expenditure
Military expenditure increase between 2021 and 2022 in the top three military expending countries and NATO, in US$ billion.
Source: SIPRI Military Expenditure Database.

presenting themselves as more environmentally friendly than they actually are. Examples of greenwashing are claiming to be on track to reduce emissions and reach net-zero, but without any credible plan presented, being non-specific and not transparent in how to reach the set goals, or using misleading labels such as 'green' that easily can be misinterpreted (UN, 2023b). Military activities like wars or preparation for wars can never be green; only some policies can reduce carbon emissions to some extent and minimise environmental impacts. Till now, the main priority has been how to adapt their bases and operations to changing climate but not doing enough to halt climate change. Given the huge amount of greenhouse gas emissions and environmental destruction produced by the military, instead of greenwashing, if NATO, is seriously concerned about climate change, it should aim at ending all the wars it fights, cutting down on the number of weapons the member countries are manufacturing and buying, set clear emission targets for the military and try all it can to manage conflicts peacefully.

The military, as a major contributor to greenhouse gas emissions and environmental destruction, plays a critical role in the global climate crisis. To genuinely demonstrate a commitment to mitigating climate change, NATO should consider strategies that go beyond superficial measures. This could involve taking decisive steps to end ongoing conflicts, which are not only sources of immense human suffering but also cause considerable environmental damage. Additionally, reducing the production and purchase of weapons by member countries can significantly lower the carbon footprint associated with military operations. Setting clear and enforceable emission targets for the military is another crucial measure. Such targets would not only contribute to global efforts to combat climate change but also set a precedent for environmental responsibility in the defence sector. Moreover, NATO's efforts should be focused on managing conflicts through peaceful means as much as possible. Emphasising diplomacy and conflict resolution can prevent the ecological and human costs of war, aligning NATO's operations more closely with the urgent need to address the environmental crisis. These steps, taken together, would represent a sincere and effective approach to integrating climate change considerations into military policy and operations.

Some apprehend that focusing on climate adaptation and mitigation efforts can affect the military's ability to defend itself, thus constraining national security (IISS, 2022). Especially if enemies are not taking measures to develop climate strategies but instead put their effort into strengthening their military capacities by continuing to use fossil fuels and emission-intense equipment. For instance, Alaska Senator Dan Sullivan has blamed the US Naval Forces for not doing enough to ensure national security as he accuses them of prioritising releasing a climate action plan instead of focusing on their shipbuilding plan to safeguard military readiness (Downing, 2023). The climate action plan emphasises the need to urgently build a climate-ready force by building climate resilience and reducing emissions (DN, 2022). The Senator argues that Russia, focusing on aggressive warfare, and China, focusing on rapid shipbuilding to conduct a military invasion of Taiwan, are the ones who are most excited about the US Navy's climate action plan as it declines the US military readiness (Downing, 2023). Currently, most military equipment available is optimised for military advantage rather than emission reduction, making it difficult to keep the same capabilities to protect the nation's security while transitioning to net-zero emissions (IISS, 2022).

However, just because climate change is not dominating global security and political discourse due to wars in the Europe and Middle East, it doesn't mean that the challenges to the health and well-being of the planet and human civilisation have disappeared. Climate change has already become a serious threat to this planet's security and survival. Highlighting how climate change is seriously affecting countries' military strength and preparation for taking serious measures to mitigate military emissions may be low-hanging fruits to elevate climate change to a security issue of top priority. However, climate change is not currently a top national security priority for any country, except for some small island states. The various ambitious actions of the large and powerful countries that are engaging in 'greening' the military are not enough to make a serious impact on reducing the effects of climate change. Firstly, the mitigation efforts to reduce emissions are not completely sincere, because climate change is not their primary concern for undergoing the necessary energy transition. Instead, military mobility, readiness, efficiency and resilience are the top drivers of reducing the dependency on fossil fuels. Secondly, a full-scale transition to renewable energy sources is complex and time-consuming, and can create new dependencies, tensions and environmental pollution. Lastly, if the military is trying to take serious steps in engaging in climate action plans for mitigation and adaptation strategies, it can be challenged by politician leaders and security analysts because it in the short term at least diminishes its capabilities to defend national security objectives. Thus, the greenwashing acts made by militaries are being challenged because the traditional roles and assignments of militaries contradict the efforts of the 'greening' of the military.

SUMMING-UP

This chapter underscores the military's significant role in contributing to global greenhouse gas emissions and environmental degradation. Military activities, from ongoing wars to the legacy of past conflicts and preparations for future wars, lead to extensive ecological damage, including deforestation, contamination and soil degradation. The production and testing of nuclear weapons are particularly noted for their lasting environmental impacts. However, there is a noted lack of transparency and accountability in reporting military emissions, with their contributions often excluded from global climate change discussions. This omission underscores the paradoxical position of the military, which, despite being a major emitter, has been reluctant to mitigate its environmental impact.

The military is a major emitter, and at the same time, the military has also become a major victim of the climate crisis. Climate change poses direct challenges to military operations, impacting bases and equipment due to phenomena like sea-level rise, extreme weather events and melting ice. These changes not only affect military readiness and the health of personnel but also extend the military's role in disaster relief operations. While NATO and other Northern militaries are starting to address climate change, efforts in the Global South lag behind. Additionally, increased military spending in response to geopolitical tensions complicates emission reduction efforts, leading to accusations of greenwashing.

The contradiction between the military's traditional roles and climate mitigation efforts is evident. The primary function of militaries, which is national defence and security, often necessitates activities and operations that are inherently emission-intensive. Efforts to reduce emissions or transition to renewable energy sources can be seen as compromising military readiness and effectiveness. This challenge is compounded by the need for militaries to maintain operational capabilities that might rely on technologies and equipment not yet compatible with stringent environmental standards. Therefore, the fundamental role of militaries in ensuring national security often stands in direct conflict with the goals of significant emission reduction and environmental protection, making it difficult for them to commit fully to climate mitigation.

While less war can lead to a decrease in military-related emissions, achieving this requires not only global political stability but also a commitment from nations to integrate environmental considerations into their defence policies and practices. This shift towards a more environmentally responsible military, in tandem with a peaceful global environment, holds significant potential for reducing the impact of military activities on climate change. Though the military is an important part of a country's framing of national security, it is not the only part as national security is much bigger than protecting the border. In this globalised world and on issues like climate change, a country as powerful as it can be, cannot really protect itself only with the help of a powerful military as it needs to negotiate and compromise in order to work together with others.

4
SHIFTING TERRITORIES AND BORDERS

This chapter explains the profound implications of climate change on the very fabric of national territories and borders. The relentless march of climate change is not just a phenomenon isolated to the polar caps; its tentacles reach as far as the majestic Himalayas, reshaping landscapes and, consequently, the geopolitical chessboard. It explores how the melting of glaciers and the rising sea-levels are not merely environmental concerns, but catalysts for potential territorial disputes and border conflicts. This chapter sheds light on the dynamic and often overlooked aspect of climate change – its power to redraw the map of the world.

Imagine a world where the lines that divide nations, the very borders that have defined countries for decades, are no longer static. This is not a scene from a science fiction novel, but a real possibility in our rapidly changing climate. It is how climate change is redrawing the map of international borders and reshaping the global battle for natural resources. Borders have long been etched into the landscape by the forces of nature. Rivers, with their meandering courses, have been the lifelines and dividing lines of civilisations. They have been sources of cooperation and conflict, shaping the destinies of nations. Mountain ranges, with their towering peaks and formidable presence, have stood as natural fortresses, marking the limits of empires. Glaciers have etched their way through these mountains, silently demarcating boundaries. And then there's the sea, vast and enigmatic, a boundary and a bridge, defining the edges of nations.

But what happens when these natural landmarks, these arbiters of borders, begin to shift? Climate change is reshaping our planet's face, redrawing borders in a way no treaty or war ever has. This chapter delves into the fascinating and alarming ways climate change is redefining territorial disputes. It is important to note how rivers, once relatively steady and reliable markers, are now capricious, their courses altered by intense rainfall patterns. Their changing shapes are not just a matter of geographical interest; they hold the potential to ignite disputes as nations scramble to hold onto or lay claim to new territories. Mountain glaciers, those ancient border guards, are retreating under the relentless sun of global warming. As they melt away, they leave behind not just altered landscapes, but questions of ownership and rights over newly exposed lands. The sea, once a symbol of permanence, is now an agent of change. Rising sea-levels threaten to swallow whole islands, reducing the territorial waters of nations and sparking new conflicts over shrinking maritime boundaries. This chapter tries to explain how nations are preparing, or failing to prepare, for these changes and how climate change is not only altering the world map, but also how it is challenging our very notions of territory, sovereignty and the rights to natural resources (Grainger and Conway, 2014).

NEW BORDER CONFLICTS ARE ON THE HORIZON

Throughout history, numerous conflicts have arisen over claims to territorial rights and the determination of political boundaries. The consequences of climate change have the potential to shift borders, thereby increasing the likelihood of new conflicts between neighbouring countries. One of the most commonly used natural markers for political boundaries is rivers. Rivers make up approximately 23 per cent of all international borders (Popelka and Smith, 2020). Globally, more than one-third of the total length of national borders is determined by rivers, as calculated by the International River Boundaries Database (Donaldson, 2009). In some cases, rivers form nearly the entire border, such as the Rio Grande, which shapes the border between the United States and Mexico, and the Senegal River, which serves as the boundary between Senegal and Mauritania. In other instances, rivers only define parts of borders. Generally, rivers have been convenient dividers in regions where European powers have exerted influence. In South America, nearly half of all borders are river-based, 28 per cent in North America, 26 per cent in Africa and 21 per cent in Europe. In Asia, rivers account for only 16 per cent of borders, where former British and French colonies play a significant role (Popelka and Smith, 2020).

Rivers can separate countries in various ways, such as by the median of the river, the centre of the main navigation channel, or the riverbank. In the latter case, the water within the river is allocated to one of the two countries, while in the other cases, the watercourse itself is divided into two parts (Lee and Tanaka, 2016). Occasionally, disputes arise over the precise location of the border along the river, which can be particularly challenging when the river's shape is changing (Grainger and Conway, 2014). For example, the border between Thailand and Lao PDR has been disputed due to disagreements about the exact points in the watershed where the border is located. In the late 1800s and early 1900s, it was agreed that the border should follow a section of the Mekong River and the Heung River. After this agreement, ownership of several villages and islands in remote mountainous areas was contested due to difficulties in precisely mapping the watershed. Variations in the river's seasonal flow and the use of maps depicting the watersheds differently complicated the border-setting process (Dommen, 1984). This border dispute led to armed conflicts and the loss of thousands of lives. In the 1980s, the case was brought to the United Nations Security Council, which decided to establish a joint technical team to determine the border. In 1996, the Thai-Lao Joint Boundary Commission was formed to clarify the boundary and settle disputes over village ownership. By 2021, 676 kilometres out of 1,810 kilometres had been demarcated (Ministry of Foreign Affairs, 2021). These kinds of border demarcations can become even more challenging with the consequences of climate change which can contribute to altering river streams.

Generally, rivers naturally alter their forms within their landscapes, growing and shrinking in water volume, eroding their beds, carrying sediment and changing their courses for a smoother flow. Climate conditions and land use changes can significantly impact how a river's form evolves, either slowly through meandering or rapidly through floods. Climate change, with more frequent and intense heavy precipitation, or prolonged droughts, can accelerate the rate and extent of river movement, and consequently, border changes (Lee and Tanaka, 2016). This is mainly because heavy precipitation and floods can increase erosion and sediment carried by the water, ultimately altering its course (Marshak, 2019). In addition to changing courses, rivers can become wider or deeper due to frequent floods (Grainger and Conway, 2014). Periods of drought can also contribute to changes in the river's course as erosion increases when plants along the riverbank dry out. Rivers can also become shallower and narrower in cases of drought or reduced precipitation. Moreover, the rising sea-level can impact the shape of the river by altering the river's baseline (Cox et al., 2022). In other words, climate change is a force that can exaggerate natural changes in the river courses and thereby risk contributing to new disputes or worsening already existing ones. Although climate change has not been pointed out as the sole or main reason why river border disputes have occurred previously or currently, it will most likely be a contributing factor that needs to be considered by political leaders.

River border disputes and their resolutions have taken various forms throughout time. In some cases, awareness of the potential need to adjust political borders defined by rivers is included in bilateral agreements, although not specifically regarding alternations due to the changing climate. For example, the border between Sweden and Finland primarily follows rivers that are prone to flooding. According to the 1809 Treaty of Fredrikshamn, the political border should be adjusted every 25 years to account for changes in the river's course. The most recent border adjustment was made in 2006, resulting in only minor changes within the watershed (Regeringskansliet, 2006). Previous border adjustments have seen islands shifting from Swedish to Finnish territory (Proposition, 1983/84:202). The bilateral agreement between Sweden and Finland serves as an example of a peaceful approach to adaptively altering the border when rivers change their course, regardless of the reason. In other cases, the process of agreeing on a new border when the previously defined one has shifted is not as straightforward. Climate-induced changes in national borders can trigger conflicts between already antagonistic neighbours or increase uncertainties between countries (Lee and Tanaka, 2016). Potentially, it can lead to conflicts in the short or long term.

Over the years, conflicts and disputes have arisen due to changes in the course of rivers that serve as national borders. One such example is the longstanding border dispute between the United States and Mexico, which lasted for nearly a century. This conflict emerged because the Rio Grande River was constantly shifting, leading to disputes over border location, property rights and the nationality of inhabitants affected by the river's movement. The dispute ultimately came to an end in the 1960s when a man-made concrete channel was constructed to control the river's course, aligning it with the agreed-upon political border. However, this resolution displaced thousands of people who found themselves on the 'wrong side' of the new artificial river border (National Park Service, n.d.). In other words, such man-made changes can significantly impact local communities, potentially forcing them to relocate. In doing so, they risk losing their properties, especially if these are situated in a country where they do not hold citizenship.

In other instances, ever-changing rivers continue to be sources of conflict. The border between India and Bangladesh is a prime example, where erosion and sediment deposition lead to the creation of new land patches and alterations in the river's form. While several rivers form the border between the two countries, border agreements exist only for the Muhuri and Fenny rivers, with their boundaries determined along the midstream of the watercourse. These rivers flow through densely populated areas where agricultural practices are common. The changing border along the river course has altered the political territory to which patches of land belong. This has led to military confrontations and land disputes between the two countries. Climate change risks deepening the already existing tension. As a consequence of the continuously changing river course and the ongoing tension, the livelihoods of people in the region have become increasingly vulnerable, uncertain and insecure.

Military interventions have also been present in a border dispute between Nicaragua and Costa Rica, whose border is shaped by the San Juan River. The border goes along the riverbank of the Costa Rican territory, giving the entire river to Nicaragua. The border area has been disputed for a long time. In 2010, a new dispute was sparked when the Nicaraguan military started dredging the river to artificially change its route. The said goal was to restore the previously existing navigable route that had been altered due to natural sedimentation processes (Lee and Tanaka, 2016). At the same time, Costa Rica argued that the dredging destabilised the hydrological system of the river which threatened ecosystems along the river basin and coast of the Caribbean Sea. The changed river course would give a substantial part of the Costa Rican territory to Nicaragua (Lee and Tanaka, 2016). Costa Rica sent around 70 police officers to protect their borders (Barquet, 2015). The case was taken to the International Court of Justice which decided that both countries should remove their forces from the conflicted area (ICJ, 2011). In 2013, the International Court of Justice decided that Nicaragua 'should refrain from any dredging and other activities in the disputed territory' (ICJ, 2013:15). The river border has been disputed over a long period both by natural river movements due to sedimentation and pollution, and through man-made changes. Climate change can further complicate this dispute as it can trigger further movement of the river and add another layer of complexity to the already intricate issue of river border disputes, affecting legal frameworks, ecological systems and the lives of those who reside along these vital waterways.

Moreover, islands situated in rivers can become subjects of disputes when the river's course changes, especially if the islands have inhabitants whose nationality depends on the river's movement. A portion of the border between Russia and China is formed by the Ussuri River. In 1969, a flood altered the river's course, shifting an island with Soviet inhabitants into Chinese territory. The Soviet Union disputed the river's shift and refused to evacuate its residents, leading to armed conflicts resulting in the deaths of hundreds of soldiers. An agreement was eventually reached in 1991, with China gaining control of over 700 islands from Russian territories (Lee and Tanaka, 2016). Similar issues involving changes in river routes and their political implications for islands have arisen in various parts of the world, including disputes between Botswana and Namibia in 1999 and Benin and Niger in 2005. These conflicts centred on territorial ownership of islands formed by rivers that define political boundaries. In such cases, the disputes mainly revolved around territorial rights to islands and border demarcation rather than water allocation between the countries (Donaldson, 2011). The International Court of Justice resolved these disputes by establishing fixed coordinates on the river but did not address how to handle potential future changes in the river's form (ICJ, 1999; ICJ, 2005).

Notably, where rivers serve as borders, there exists a potential risk of border conflicts unless substantial measures are taken to address the challenges arising from shifting rivers and altered political boundaries. Conflict resolution has led to three approaches for dealing with changing borders: continuous adjustment of the border in response to natural river movement, stabilising the riverbed through the construction of a concrete channel or fixing the border by coordinates (Donaldson, 2011). Many river border disputes remain unresolved, and the influence of climate change can exacerbate existing border conflicts or initiate new ones due to its significant impact on river forms and political boundaries. Additionally, the altering course of border rivers can significantly impact the livelihoods of farmers and fishing communities, who may face the loss of agricultural land and fishing rights.

GLACIAL BOUNDARIES ARE SHIFTING

The accelerated melting of glaciers as a consequence of climate change can contribute to escalating border conflicts. Glaciers, which cover about ten per cent of the Earth's surface, provide water for nearly two billion people and serve as the largest freshwater reservoir globally (UN, 2023c). They play a crucial role in the global climate system by absorbing greenhouse gases and redistributing heat. However, climate change is causing glaciers to rapidly retreat, and when this process begins, the melting of glaciers can accelerate due to changes in their structure. Additionally, glaciers help regulate atmospheric temperatures because their reflective white surfaces bounce the sun's rays back into space, rather than absorbing heat. In simple terms, as glaciers melt, fewer ice and snow-covered areas remain, and the exposed darker surfaces absorb more heat from the sun, further accelerating glacial melting (Marshak, 2019). Glaciers are vital for water and food supplies and are crucial for thriving ecosystems in mountainous regions. Rapid glacier melting can have devastating effects on local communities and ecosystems.

Naturally, glaciers move over time as they retreat and advance, depending on the balance between snow accumulation and glacial melt. However, accelerated glacial melting caused by climate change significantly alters local topographies and can affect geopolitical relations. Glacial melting also contributes to rising sea-levels. Melting glaciers can lead to changes in borders by altering the form of rivers or reshaping the topography in areas where glaciers are part of the border. Melting glacier runoff flows into rivers, some of which define borders between countries. Climate change-induced rapid glacial melting can increase the likelihood of flooding, leading to more frequent reshaping of rivers. Ultimately, the loss of glacier mass can cause rivers to dry up, affecting border areas. Similarly, political borders based on mountain ranges may change as glaciers retreat or advance if borders cross them. Mountain ranges have often been used as political boundaries because they are difficult to traverse and have been considered natural dividers between nations. When glaciers form part of political borders, the highest point of the glacier is often used to divide neighbouring countries. Depending on the area's topography, the glacier's highest point can mark the starting point for two separate watersheds and be considered a natural divider. In other cases, steep glacier slopes are used as dividers (Lee and Tanaka, 2016). However, the remote and inaccessible locations of glaciers and high mountain ranges have often made it challenging to precisely map border locations, contributing to border disputes when neighbouring countries claim different territorial rights. Accelerated glacial melting can exacerbate uncertainties regarding border locations, leading to increased tension and heightened disputes. Many of the glaciers that form borders are located in the Alps, the Himalayas and the Andes. All three regions have experienced border disputes involving glaciers located in border areas.

Glaciers in the Alps are highly susceptible to climate change. The Swiss Alps experienced a record mass loss in 2022, with a ten per cent reduction in ice volume between 2021 and 2023 (WMO, 2023b). Since 2001, the Swiss Alps have lost over one-third of their glacial mass, decreasing from 77 km 3 to 49 km 3 in 2022. The reduction in glacial mass is a consequence of below-average winter snowfall and extended, persistent summer heatwaves. In 2022, dust from the Sahara darkened the glacial surface, accelerating ice melt (WMO, 2023a). The melting glaciers in the Swiss Alps have geopolitical implications, as the border between Switzerland and Italy partly consists of glaciers. The watershed beneath the glacier marks the border (Ferrari et al., 2018). As glaciers recede, they alter the position of the watershed and, consequently, the border's shape. In 2009, the two countries established a commission to review their borders every few years as the glaciers retreat (Lee and Tanaka, 2016). Since then, the border between the countries has been legally defined and accepted as a moving border that will change over time depending on glacier changes. While Switzerland shares glacial borders only with Italy, Italy shares glacial borders with Austria and France as well. The border between Italy and Austria has a similar agreement to the one between Italy and Switzerland, defining the border as a moving one reviewed every few years. This agreement was established in 2006 (Ferrari et al., 2018). However, the border between Italy and France does not have such an agreement. Instead, the border is contested as the two countries claim different territorial rights over the Mont Blanc mountain and glacier. This border dispute dates back to the 1800s when the Treaty of Turin declared the border to be on the summit of Mount Blanc. However, France later changed the border's location to resolve an internal territorial dispute between two French municipalities (Iannizzi, 2021). The border dispute between France and Italy is not caused by glacier movement, but it remains ongoing, and there is no agreement on how to address the retreating glaciers in the border area.

Melting glaciers risk intensifying pre-existing border disputes (Grainger and Conway, 2014). In the Andes, approximately 4,000 glaciers are situated along the border between Chile and Argentina (Schoolmeester, 2018). Most border disputes between the two countries have been resolved. However, one dispute remains unsettled, specifically concerning the location of the border in connection to the Southern Patagonian Ice Field in Chile or the Continental Ice in Argentina. The ice field is the largest water reserve in the southern hemisphere, yet demarcation has not yet been determined (Manzano Iturra, 2019). It is a vital water resource for both countries and holds strategic significance due to its location. This glacier is the fastest-melting worldwide and significantly contributes to sea-level rise. Along with the Northern Patagonian Ice Field, it contributed to three per cent of sea-level rise between 2002 and 2017, making it one of the largest glacial contributors to sea-level rise per unit area globally. Several attempts have been made to define the border between the two countries, but differences in definitions and maps depicting the ice field area have left the border disputed (Manzano Iturra, 2019). To prevent an armed border conflict, the presidents of both countries signed an accord in 1998 to divide the ice field (L.A. Times Archives, 1998). The accord stipulated that a Joint Commission, consisting of members from both countries, would finalise the demarcation on a 1:50,000 scale map, where high peaks dividing waters would determine borders (Vivar, 2020). To date, no final demarcation has been agreed upon, and the issue becomes more challenging as glaciers retreat, altering high peaks and watersheds, similar to the case in the Alps.

Glaciers also move rapidly in the Himalayas. After Antarctica and the Arctic, the Himalayan Mountains possess the largest snow and ice deposits globally. The cryosphere, part of the Earth covered by water in solid form (e.g., ice and snow), in the Himalayas is often referred to as the world's water tower, estimated to directly supply water to nearly 1.5 billion people (National Geographic Society, 2022a). The Himalayas are warming at a rate faster than any other region, except for Antarctica and the Arctic (Panday, 2021). Changing climate conditions are causing the Himalayan glaciers to shrink in size and retreat. A recent study by various government agencies and research institutes in India found that the Hindu Kush Himalayan glaciers are retreating by nearly 15 metres annually, with less retreat in the Karakoram region (PIB Delhi, 2022). Several studies confirm that climate change has caused alarming glacier melting rates in Nepal's Himalayan region (Wester et al., 2019; Wood et al., 2020). The melting of Himalayan glaciers has doubled in the last two decades (Maurer et al., 2019). Satellite images reveal that the Himalayas are losing eight billion tonnes of ice annually, and this loss is not being replenished by new snowfall. The glacier surface diminished by an average of 22 cm annually from 1975 to 2000, but the average annual loss increased to 43 cm from 2000 to 2016 (Wester et al., 2019). Unless climate change is reversed, at least one-third of the ice in the Himalayan range is expected to melt by the end of this century (Panday, 2021). Human-caused climate change is the primary cause of accelerated glacier melting, but high-altitude desert dust and air pollution also contribute to this phenomenon. Mountain ranges in the Himalayas serve as borders between several countries in the region, and many of these borders are contested. The melting glaciers risk intensifying border disputes in the region and addressing this emerging challenge may not be as smooth as the agreements between Italy and Switzerland.

For example, the border between India and Pakistan is disputed over the Siachen Glacier in the Karakoram range. The glacier has been a source of contention for decades. Due to its inaccessibility, the glacier had not been previously mapped. India has held control over the glacier since 1984 (Joshi, 2017). Currently, both countries claim sovereignty over the entire region where the Siachen Glacier is located. This border dispute has led to numerous casualties and injuries, earning the glacier the title of the world's highest battlefield (Swain, 2009). However, most deaths and injuries have been caused by harsh weather conditions. In 2012, an avalanche killed 140 Pakistani soldiers, and over the years, nearly 900 Indian soldiers have perished while trying to control the area. Moreover, the militarisation and increased human presence in the region contribute to accelerated glacial melt (Orlove et al., 2008). Alternative solutions to the border conflict have been proposed, such as converting the Siachen Glacier into a transboundary peace park to build trust between the two countries (Swain, 2009). Although a ceasefire has been in place since 2003, a resolution to the conflict remains elusive (Joshi, 2017). Border disputes also exist between India and China, with the world's longest disputed border. The Himalayan border region between China and Nepal is also disputed. The China-Nepal border spans 1,400 km along the Himalayan Mountains, with many remote areas making it challenging to precisely determine the border's location. Additionally, the rapidly melting glaciers are expanding the transboundary lake catchment shared by both countries (Zhong et al., 2021). Changing geological conditions in the area poses significant risks to regional and human security. In the Himalayan region, pastoralist communities are at risk of being impacted by border changes, as the land they utilise for grazing could become the territory of another country.

Although glacial border disputes are found in different parts of the world, it is primarily the case in the Alps that can be observed as a dispute stemming from melting glaciers, partly connected with global warming. Other glacier-related border disputes mainly arise from unresolved territorial claims rather than rapid glacier melting. Nonetheless, as topographies in these areas change, uncertainties regarding border locations can increase. In other words, melting glaciers can serve as triggers for heightened tensions between neighbouring states.

As glaciers melt, they dramatically alter the topography of border regions. This can lead to uncertainties regarding the exact location of borders, especially in rugged, mountainous areas where glaciers have traditionally marked boundaries. Such uncertainties can strain relationships between neighbouring states, especially if the areas in question hold strategic or economic value. Melting glaciers can act as catalysts for increased tensions between neighbouring states. As glaciers recede, new land is exposed, potentially leading to competition for control or access to these new areas. This is particularly relevant for regions rich in natural resources, where new land could mean access to valuable minerals, water sources or strategic military positions. Given the potential for conflict, there is a growing need for proactive diplomacy and international cooperation in managing glacier-related border disputes. This might involve revising existing treaties, engaging in joint monitoring of glaciers and developing mechanisms to peacefully resolve disputes as they arise due to changing landscapes.

MELTING SNOW IS DISSOLVING POLITICAL STABILITY

The melting of glaciers in mountain ranges isn't the only factor that can lead to territorial disputes; the melting sea ice in the Arctic Ocean can also contribute to geopolitical challenges and competing claims. The region north of the Arctic Circle is predominantly covered in ice, holding almost 20 per cent of the world's freshwater resources (National Geographic Society, 2022b). Arctic ice sheets play a crucial role in regulating the planet's ocean and atmospheric temperatures by cooling the water and reflecting sunlight back into space. Recent studies indicate that climate change is causing the Arctic to warm up four times faster than the global average since 1979 (Rantanen et al., 2022). Over the past few decades, the thickness of Arctic ice during the summer months has decreased by 65 per cent. This trend is expected to result in the Arctic being ice-free in the summer months as early as 2030 (Kim et al., 2023) and possibly even ice-free in the winter months by 2050 if greenhouse gas emissions remain high (Notz and SIMIP Community, 2020).

The warming in other parts of the world also affects the Arctic. Warmer water from the Atlantic Ocean is reaching deep into the Arctic region, leading to a mixture of warm and cold waters. This warm water contributes to the acceleration of ice melting. The dwindling sea ice and warmer temperatures have severe consequences for Arctic ecosystems and the traditional ways of life for indigenous people in the region. The Arctic is home to approximately four million people, with nearly 10 per cent of them belonging to indigenous communities (IPCC, 2019b). The Arctic region is divided among eight countries with territorial rights within the area. These Arctic states include the United States, Canada, Iceland, Denmark (including Greenland), Norway, Sweden, Finland and Russia. In 1996, these countries established the Arctic Council to promote cooperation on environmental and sustainable economic development issues in the region. However, the council's mandate is limited and lacks legally binding power.

For centuries, the harsh climate and frigid temperatures had kept the Arctic largely untapped and unexploited. Yet, in recent decades, global warming has warmed the Arctic, intensifying competition among global powers to access and control this largely uncharted and resource-rich region. The rapid melting of ice caps has opened numerous opportunities for the world's divided nations, potentially leading to new geopolitical tensions and conflicts. Under the receding ice, the Arctic region boasts extensive reserves of oil, natural gas and minerals. Rough estimates suggest that the area holds around 13 per cent (equivalent to 90 billion barrels) of the world's oil, surpassing

Russia's and the United States oil reserves combined (Thangaraj and Chowdhury, 2022). Additionally, it's believed to house 30 per cent (approximately 48 trillion cubic metres) of the world's undiscovered natural gas reserves, as well as significant deposits of copper, lithium, nickel, platinum and rare earth metals. The easier-to-access minerals and metals in the rapidly thawing Arctic region are estimated to be worth approximately US$1 trillion (National Intelligence Council, 2021). These natural resources are essential for renewable energy technologies. The region also offers substantial potential for generating renewable energy through wind and marine turbines, critical for countries aiming to meet their energy needs while reducing emissions. For example, the European Union views the Arctic as a promising site for boosting renewable energy production, addressing energy crises and tackling climate change simultaneously, reducing dependency on non-EU countries like Russia and China in pursuit of a 'green transition' (European Commission, 2021). Yet, despite concerns about climate change, oil and natural gas production in the Arctic, particularly in Russia, Norway and the United States, is on the rise. The Norwegian Parliament has also approved opening up for deep-sea mining in its part of the Arctic Ocean (Paddison, 2024).

Beyond providing access to previously inaccessible natural resources, the receding sea ice is opening new shipping routes through the Arctic Ocean. The two primary routes within the Arctic Circle are the Northern Sea Route along the Russian coast and the Northwest Passage through the Canadian Arctic Archipelago. These routes are experiencing more frequent ice-free conditions in the summer months. With the ongoing decrease in sea ice, these two routes could become more reliable throughout a significant portion of the year. The former US Secretary of State Michael Pompeo referred to the two routes in the Arctic as the '21st Century Suez and Panama Canals' in 2019 (Pompeo, 2019). The Northwest Passage connects the Atlantic and Pacific Oceans and is approximately 9,000 kilometres shorter than the route through the Panama Canal and 17,000 kilometres shorter than the journey around South America's Cape Horn (NOAA, 2016). The Northern Sea Route could be almost 40 per cent shorter and 10–15 days faster than the current route through the Suez Canal for shipping goods between China and Europe (Arctic Review, 2022). In 2022, despite global turmoil and sanctions, the Northern Sea Route experienced increased traffic, driven by the Russian Vostok Oil Project's initiation, which is expected to further boost traffic from 2024 (Humpert, 2023).

Access to natural resources and shipping routes in the Arctic region is disputed due to overlapping national claims in the Arctic Ocean. The United Nations Convention on the Law of the Sea (UNCLOS), adopted in 1982, was designed to resolve disputes over the rights to use the sea and its resources. UNCLOS introduced the concept of Exclusive Economic Zones (EEZ), extending up to 200 nautical miles (370.4 kilometres) beyond a nation's land boundary. Within the EEZ, the country has sovereign rights over the living and non-living natural resources in the sea, subsoil and ocean floor. Additionally, countries can claim an extended continental shelf under certain conditions, allowing them sovereign rights for resource exploitation within the seabed and subsoil, such as oil and natural gas. Fish stocks are excluded from this sovereign right, distinguishing it from the EEZ. The legal definition of the continental shelf encompasses the submerged 'seabed and subsoil of the submarine areas that extend beyond its territorial sea throughout the natural prolongation of its land territory to the outer edge of the continental margin, or to a distance of 200 nautical miles from the baseline from which the breadth of the territorial sea is measured where the outer edge of the continental margin does not extend up to that distance' (UNCLOS, 1982:53). The continental shelf represents an underwater extension of a country's land territory, and its outer edge can be determined using two different complex methods: measuring the thickness of sedimentary rocks in the soil or measuring a fixed distance of 60 miles from the base of the shelf's slope (CLCS, 2012). Continental shelves often contain significant natural resources, including oil, natural gas and minerals (Fletcher Primer, 2023). The Arctic Ocean is particularly shallow, featuring extensive continental shelves. All coastal nations within the Arctic region have claimed continental shelves extending from their EEZs, including the United States, Canada, Russia, Norway, Denmark (Greenland) and Iceland. Only a small portion of the Arctic Ocean remains unclaimed. These claims are submitted to the UN Commission on the Limits of the Continental Shelf. Although the border between the United States and Russia's territories is agreed upon, many other claims are contested, primarily due to challenges in measuring continental shelves in the icy environment. Some claims even overlap. For instance, the Lomonosov Ridge, which crosses the North Pole, has been claimed by Russia, Denmark and Canada (as shown in Map 4.1). Denmark claims it as an extension of Greenland, Russia claims it as an extension of the Siberian archipelago Franz Josef Land and Canada claims it as an extension of Ellesmere Island within their territory (BBC, 2014; Henriques, 2020). Another example of contested water rights is the Northwest Passage, which Canada claims to be internal waters under their full sovereignty, while the United States asserts that it lies within international waters, allowing for the right of innocent passage by all countries (Tømmerbakke, 2019). In fact, the former US Secretary of State suggested that Canada's claim to the Northwest Passage is 'illegitimate' (Pompeo, 2019). As natural resources and more accessible shipping routes become available in these disputed areas, the risk of heightened tensions among Arctic states increases. The increasing tensions can become particularly challenging for the indigenous populations and local communities in the area who to various extent are dependent on the local environment for their livelihoods or traditional practices. There is a prominent risk that the indigenous and local interests are not sufficiently taken into account when the exploration and exploitation of natural resources in the region increases. Ultimately, some indigenous groups might risk losing what they have left of their traditional practices, and risk being displaced by exploitation projects. Indigenous rights are protected by the International Labour

Organization (ILO) 169 convention. ILO 169 recognises indigenous peoples' connection with their traditional land territory, gives them the right to preserve and develop traditional indigenous practices and should participate in decision-making processes in the area where they live (ILO, 1989). Out of all the countries in the Arctic region, only Norway and Denmark have ratified the convention (ILO, 2024). However, even Norway and Denmark have been internationally criticised for not fully respecting indigenous rights. When the Arctic is becoming an increasingly interesting area for exploitation, it is the indigenous people in the region that are at most risk of being harmed as the interests of exploiters may stand in conflict with the interests of the indigenous groups.

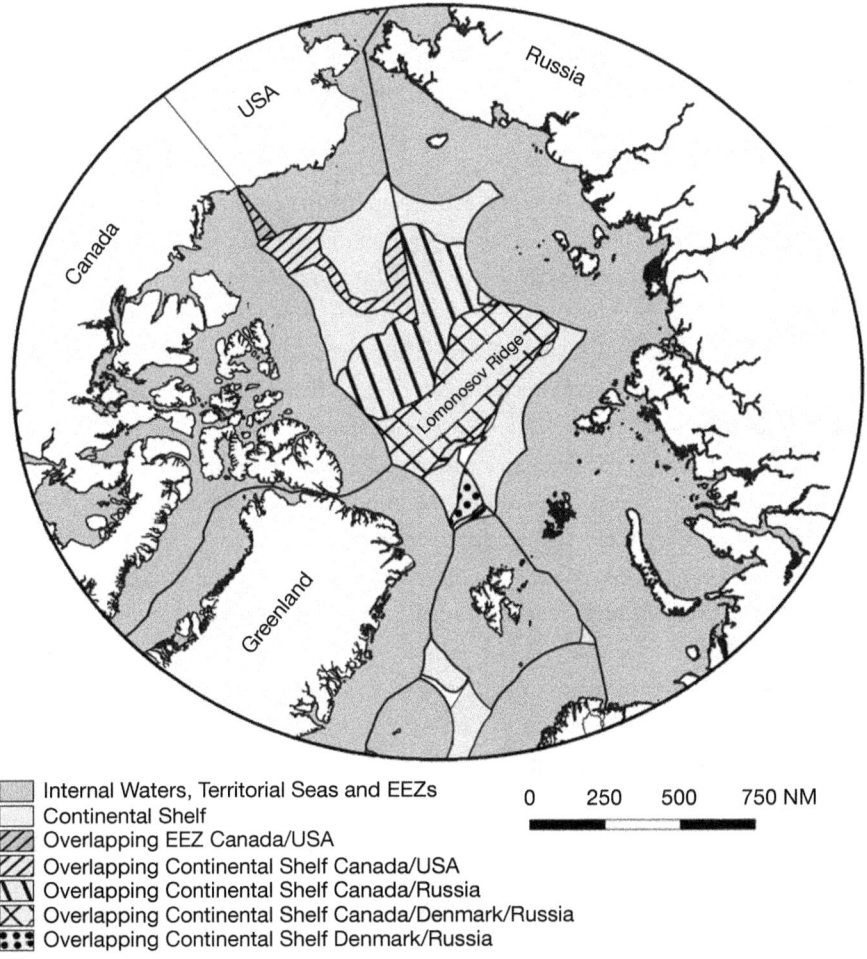

Map 4.1 Overlapping Claims in the Arctic Ocean

Note: The map shows the various overlapping claims of continental shelves and EEZs in the Arctic Ocean. The map is visualised by Maria Båld.

At the same time, the emerging security challenges in the Arctic region have significant geopolitical consequences, and the current framework for cooperation through the Arctic Council may not be sufficient to address them. Military security concerns have been explicitly excluded from the Council's purview due to the sensitive nature of the topic (Arctic Council, 1996). The Russian invasion of Ukraine disrupted cooperative efforts in the Arctic Council since Russia was chairing the Council at the time. Russia was subsequently suspended from the Council, and its return remains uncertain (DoS, 2022). Besides the member states, the Arctic Council includes 38 observers, comprising many major European countries, China, India, Japan, South Korea, Singapore and various intergovernmental, interparliamentary and non-governmental organisations (NGOs). Non-Arctic nations are increasingly cooperating with and investing in Arctic projects. China, France, India, Japan, South Korea and the United Kingdom have all released their own Arctic strategies, primarily focusing on economic opportunities and security concerns (National Intelligence Council, 2021). China, in particular, has gained importance in various aspects of Arctic affairs, from exploring natural resources to northern shipping and environmental protection. China considers itself a 'near-Arctic State' and has significantly expanded its engagement in the region, aiming to establish a substantial geostrategic presence (Dams et al., 2020). China has sent numerous scientific expeditions to the Arctic and plans to build nuclear-powered icebreakers to connect Arctic routes to its Belt and Road Initiative (Humpert, 2021). As part of its global power aspirations, China's navy seeks faster access to the Atlantic Ocean, which could be facilitated by the Northern Sea Route. The Chinese Communist Party has even proposed that Arctic resources be shared among all countries based on population size (Doshi et al., 2021). Despite initial Russian opposition to China's Arctic Council membership in 2007, their cooperation has grown significantly following Western sanctions against Russia since 2014. China has provided financial support for various Russian Arctic projects, including the Yamal Liquified Natural Gas project. In return, China gains shorter trade routes and increased access to the Arctic's natural resources. China's growing involvement in Arctic geopolitics reduces the potential for Russia's exclusion or isolation in the region. It's only a matter of time before the Arctic becomes another arena for power competition between the world's major players in an increasingly bipolar world. Ultimately, this will have critical effects on achieving justice and equality in the region.

Russia has already reactivated more than 50 Soviet-era military outposts in the Arctic region and strengthened its navy's northern command, responsible for the Arctic, by deploying advanced warships and testing its Arctic-based military capabilities. In response, the United States is deploying fifth-generation fighter jets in the Arctic, increasing submarine patrols and, along with its NATO allies, expressing greater concerns over Russia's expanding military activities, especially in light of the Ukraine conflict (Hadley, 2023). Moreover, following Finland's and Sweden's recent NATO memberships, Russia will be the sole non-NATO member among the Arctic states. Russia and NATO regularly conduct large-scale military exercises in the Arctic. In 2018, NATO conducted one of its largest exercises, Trident Juncture, in northern Norway, while Russia conducted Umka-2021, and NATO responded with the 'Cold Response' in 2022. Under the Biden administration, the United States is substantially investing in enhancing its military capabilities in the Arctic. The US's 2022 national strategy for the Arctic emphasises strengthening security to defend the United States and allies' interests in the region as the top priority (The White House, 2022). Simultaneously, Russia is becoming increasingly assertive in its interactions with other Arctic nations, leading NATO countries to bolster their military presence and exercises in the region, further escalating tensions.

In conclusion, climate change has introduced unprecedented security challenges to the Arctic region, driven by competing interests in access to resources, increased military presence and the opening of new shipping routes, all exacerbated by global warming. These developments pose a threat to fragile Arctic ecosystems and the livelihoods of indigenous populations, who face the risk of displacement and loss of traditional ways of life. Geopolitical tensions in the Arctic are on the rise, and the current mechanisms for cooperation and conflict resolution may not be adequate to address the emerging challenges. The Arctic region is becoming a focal point of global power competition, with multiple countries, including China, vying for influence and access to its abundant resources.

China's pursuit of a 'Polar Silk Road' reflects its ambition to play a significant role in the Arctic, further complicating the geopolitical landscape. As the Arctic ice continues to melt and new opportunities emerge, it is increasingly likely that the region will become a centre of geopolitical confrontation. This evolving scenario may not only involve Arctic states but also other major powers seeking to expand their influence, access resources and capitalise on new shipping routes. Climate change is reshaping the Arctic in profound ways, not just environmentally but also geopolitically. The combination of resource competition, military build-up, environmental concerns and the involvement of global powers like China is creating a complex and potentially volatile situation. It highlights the need for robust, adaptive and forward-thinking international policies to manage these emerging challenges in one of the world's most sensitive and strategic regions.

DISPUTES OVER ISLAND BORDERS

One potential cause of territorial disputes triggered by climate change is the rising sea-level, which is reducing the size of land and islands. This phenomenon can have significant implications for countries' sovereign rights over sea areas.

The Intergovernmental Panel on Climate Change (IPCC) estimates that if high emission levels continue, global sea levels could rise by approximately 1.1 metres by the year 2100 (IPCC, 2019b). This unprecedented sea-level rise is primarily driven by the accelerated melting of glaciers and ice sheets, as well as the warming of seawater. Additionally, excessive groundwater pumping in various parts of the world is contributing to this rise, although it's considered a relatively minor factor (Wada et al., 2016a). Notably, the loss of ice in Antarctica and Greenland accounts for one-third of the global sea-level rise. The rate of ice loss in these areas has increased sixfold between 1992 and 2018, aligning with the worst-case climate warming scenario outlined by the IPCC. If this trend continues, it could result in an additional 17 cm of sea-level rise by 2100 (the IMBIE Team, 2018; 2019). However, the melting of sea ice in the Arctic, while concerning for other reasons, contributes minimally to global sea-level rise because the melted sea ice turns into seawater. The volume of water from the melted sea ice is roughly equivalent to the volume of water displaced by the ice. Nonetheless, when seawater warms, it expands, and the accumulation of greenhouse gases in the atmosphere, primarily due to human activities, has raised the planet's temperature. Approximately 90 per cent of this trapped heat is absorbed by the world's five oceans (NASA, 2023b). Consequently, the warming oceans account for around one-third of the global sea-level rise (NASA, n.d.). As global temperatures continue to increase, sea levels are expected to rise further.

The rising sea-level is a significant concern for global security, as it can exacerbate tensions between countries. In February 2023, the United Nations Security Council held a high-level open debate on sea-level rise and its implications for international peace and security. The UN Secretary-General described the rising sea as a 'threat multiplier' and noted that its impact is already creating new sources of instability and conflict (UN Secretary-General, 2023). Small island nations, particularly those in the Pacific and Indian Oceans, have immediate concerns related to the shrinking size of their remote islands, which could reduce their EEZs. These zones grant countries sovereign rights over natural resources. The submergence of islands raises complex legal issues that could trigger territorial disputes (Freestone and Cicek, 2021). The UN Convention on the Law of the Sea (UNCLOS), adopted in 1982, did not account for such legal challenges associated with rising sea-levels, and this lack of clarity over future rights to fish and other natural resources in contested areas is a growing concern. The primary issue pertains to the reduced EEZs for countries that heavily rely on the fishing industry.

EEZs are defined as sea territories that extend up to 200 nautical miles (370.4 kilometres) from a country's defined shoreline baseline. Within these zones, countries hold sovereign rights over the living and non-living natural resources found in the water, subsoil and seabed. While the implications of sea-level rise on EEZs have not yet been legally tested, the existing definitions in UNCLOS, along with past territorial disputes, provide insights into potential challenges. The central concern revolves around the rights associated with islands. UNCLOS defines an island as 'a naturally formed area of land, surrounded by water, which is above water at high tide' (article 121, UNCLOS, 1982). UNCLOS distinguishes between islands and rocks, with rocks being defined as landforms that 'cannot sustain human habitation or economic life of their own' (article 121, UNCLOS, 1982). Islands have legal rights to EEZs, while rocks do not possess EEZs or continental shelves. In essence, if an island becomes submerged to the point where it is no longer habitable, the right to fish in the waters surrounding that island may be significantly reduced, or even eliminated (Moritaka, 2013). This is a pressing concern for island nations consisting of numerous small islands scattered across the ocean, as each habitable island is surrounded by its own EEZ.

The distinction between rights associated with rocks and islands has already led to disputes between neighbouring countries. For instance, China, Taiwan and Japan have an ongoing dispute regarding their overlapping EEZs, particularly concerning small atoll islands. Japan claims that these islands are indeed islands, thereby encompassing them in its EEZ. China and Taiwan, on the other hand, argue that the landforms are rocks and should not be part of Japan's EEZ. This dispute centres on areas with significant natural resources, including fish stocks, natural gas and oil (ICWA, 2023). Another example involves the delimitation of overlapping EEZs, which was recently determined by the International Tribunal of the Law of the Sea (ITLOS) between Mauritius and the Maldives. The dispute revolved around the Chagos Islands, whose EEZ overlapped with the EEZ of the Maldives. The ownership of the Chagos Islands has been disputed for decades between Mauritius and the United Kingdom. This has previously complicated the delamination process of the overlapping EEZs, considering that it was unclear which country the Maldives should negotiate with. In the 1960s and 1970s, the United Kingdom expelled the islanders and established a UK/US military base in the Chagos Archipelago. In 2019, the ICJ's advisory opinion was that the United Kingdom had not lawfully completed its decolonisation of the Chagos Islands and should return the islands to Mauritius (Mujuthaba and Brewster, 2021). Currently, there is an ongoing case about the sovereignty of the islands and resettlement for the population descending from the Chagos Islands through their right to permanently return to their home islands (UK Parliament, 2024). After the ICJ advisory opinion, Mauritius claimed a larger part of the EEZ north of the Chagos Islands, which overlapped with the Maldivian EEZ. The case was brought to the ITLOS, and finalised in 2023. The contestation concerned the baseline from which nautical miles should be measured and highlighted the complexities of resolving disputes related to territories consisting of islands with ambiguous definitions under UNCLOS (Quell, 2023). The ruling was in favour of the Maldives, given a slightly larger share of the contested 92,653 km^2 (ITLOS,

2023). However, previously the whole disputed area belonged to the Maldives. Therefore, local fishermen in the South of Maldives expressed unhappiness as a part of an area rich in fish had been lost to Mauritius (Maldives Financial Review, 2023). While these particular disputes do not directly result from sea-level rise, they illustrate how disagreements can arise over the rights to fish and explore natural resources in territories with islands of disputed definitions. As sea levels continue to rise and partially or completely submerge islands, the risk of future conflicts and disputes over rights to ocean areas and natural resources becomes increasingly prominent.

The issue of disappearing islands has led to reduced EEZs in the past. The United States and Mexico, for instance, contested the boundaries of their adjoining EEZs in the Gulf of Mexico in the early 2000s when a Mexican island could no longer be located. This island, first mapped in 1539, played a crucial role in determining the outer limits of Mexico's EEZ. After an extensive search and examination of the ocean floor, the National Autonomous University of Mexico (2009) concluded that the island likely never existed in the documented location. This discovery significantly reduced Mexico's EEZ in an area rich in oil resources (Collins, 2021). Although this case is unique because it raises the possibility that the island may have never existed, it highlights the uncertainty surrounding EEZs of islands that may become submerged due to rising sea-levels.

To safeguard their sovereign rights over natural resources, countries in the Western and Central Pacific region have attempted to establish permanent boundaries for their existing EEZs. Permanent EEZ boundaries are seen as a means to protect food security, sustenance, exports and industry, particularly for countries heavily reliant on the fishing sector (Phillips, 2022b). Ten small island states in the Pacific have a combined land area roughly the size of Florida, but their collective EEZs are larger than the surface of the moon. This is mainly due to the presence of several remote islands, each of which is surrounded by its own EEZ (Doyle, 2021). The Pacific Ocean is abundant in fish, particularly tuna, with the western and central Pacific Ocean accounting for approximately 55 per cent of global tuna production. Half of this catch is sourced from the EEZs of Pacific Island countries (WCPFC, 2022). Reduced EEZs can lead to devastating socio-economic and cultural impacts on affected populations, particularly when compounded by the warming of seas and increased ocean acidification that can alter fish migration patterns. For example, it is estimated that by 2050, the total catch of three tuna species could decline by an average of 13 per cent in the combined EEZ of ten small island states in the Western Pacific region, primarily because these fish are expected to move to international waters (Bell et al., 2021). Changes in fish stocks have broader implications for ecosystems, biodiversity and livelihoods, and can even trigger geopolitical tensions.

In 2021, the Pacific Island Forum issued a joint declaration on preserving maritime zones in light of sea-level rise due to climate change. The declaration underscores the commitment of its 18 member states to maintain their maritime zones without reduction, without reviewing or updating the baselines and outer limits of their maritime zones, even if they are affected by rising sea-levels. The declaration also emphasises principles of equity and justice, recognising that the countries most impacted by sea-level rise are among those that have contributed the least to it (Pacific Island Forum, 2021). However, this effort to establish permanent EEZ boundaries in the Pacific lacks a legal mandate under UNCLOS, and renegotiating UNCLOS will be a challenging task. Many non-Pacific countries stand to benefit from increased access to international ocean areas in the region, even as Pacific countries face uncertainties related to their economic rights (Phillips, 2022b). In essence, sea-level rise introduces substantial economic uncertainty for small island nations and carries profound geopolitical implications, increasing the likelihood of territorial conflicts. Sea-level rise is no longer a hypothetical concern; it has become a harsh reality that requires global attention and comprehensive solutions.

The rising sea-level, driven by climate change, poses a significant threat to territorial integrity and sovereignty, particularly for small island nations. This phenomenon threatens to reduce the size of land and islands, impacting countries' EEZs and sovereign rights over sea areas. Already, disputes have arisen over the rights associated with islands and rocks, with countries contesting the definition and entitlements of these landforms. As sea levels continue to rise, partially or completely submerging islands, the risk of future conflicts over ocean areas and natural resources becomes increasingly prominent. The rising sea-level is thus not only an environmental concern but also a growing geopolitical issue, necessitating global attention and comprehensive solutions.

SUMMING-UP

The escalating effects of climate change on global security demand urgent attention and action from the nation-states and the international community. Alterations in river courses, retreating glaciers, melting Arctic ice sheets and rising sea-levels are reshaping political borders and heightening national security concerns. These changes are not confined to mere territorial disputes but extend into broader spectrums of geopolitical instability and resource competition. Rivers, traditionally used as natural markers for political boundaries, are now subject to the whims of changing climate patterns. This is exemplified by disputes over rivers like the Mekong and the challenges faced by countries such as Thailand, Lao PDR, India and Bangladesh. Human interventions, like dredging activities in the case of Nicaragua and Costa Rica, further compound these tensions.

Glacial melting presents another grave challenge. As glaciers retreat, they alter local topographies and existing political boundaries, especially in regions like the Alps, the Himalayas and the Andes. This poses risks to regional security and the livelihoods of communities reliant on these natural resources. The melting glaciers between Switzerland, Italy, Austria and France, and the disputes over the Siachen Glacier between India and Pakistan, are stark reminders of the emerging security threats. In the Arctic, the rapid environmental transformation due to climate change is unveiling new pathways for resource exploitation and shipping, drawing attention from global powers and intensifying competition. The Arctic, with its rich reserves of oil, natural gas and minerals, is becoming a focal point of global power dynamics. This scramble for resources and strategic advantages is testing the limits of existing legal frameworks and escalating military presence in the region, as seen with Russia's activities and NATO's responses. At the same time, it threatens the Arctic ecosystems and indigenous and local communities' livelihoods in the region.

Furthermore, rising sea-levels threaten the territorial integrity and sovereignty of small island nations, potentially leading to the loss of EEZs and ensuing legal and territorial disputes. This creates not only environmental but also socio-economic and cultural challenges for these nations, particularly those heavily reliant on fishing. To effectively address these multifaceted security challenges, a coordinated global response is essential. This involves adapting existing legal frameworks, fostering international cooperation, and developing robust strategies to manage the geopolitical landscape reshaped by climate change. The international community, particularly rich and powerful countries, must act decisively to mitigate these risks, ensuring peace and stability in a rapidly changing world.

5
WATER CONFLICT AND COOPERATION

As the world grapples with the escalating impacts of climate change, one of the most critical arenas of concern is the effect on water – a vital, shared natural resource. This chapter delves into the complex interplay between climate change and water, uncovering how alterations in the water cycle are precipitating a spectrum of challenges: from water scarcity to floods, and from pollution to the threat of hydro-project destruction. The United Nations Environment Programme (UNEP) has sounded the alarm, pinpointing the water crisis as a top global risk in the coming decade. This crisis is not just about human survival; it's a linchpin for thriving ecosystems and economic prosperity. This chapter illuminates the multifaceted implications of water as a focal point of climate-induced stress, exploring how it's becoming a nucleus for resource conflicts, challenging existing water agreements and sparking intense debates over the potential for 'water wars.' Simultaneously, it examines the prospects of water as a conduit for cooperation, underscoring the urgent need for innovative water-sharing arrangements and sustainable management in the face of a changing climate.

WATER AND CLIMATE CHANGE

The most profound effect of climate change is on the water cycle (Earle et al., 2015). Mainstream debates about the impacts of climate change on the water system have mainly revolved around glacier melting and sea-level rise. Less attention has been given to the impacts on rivers, lakes and aquifers that are significant sources of freshwater supply for human use. These freshwater systems are going through major changes worldwide due to warming temperatures. Mainly, global warming changes the predictability of the water cycle. It alters the water cycle through changed evaporation and soil moisture systems, as well as altered intensity, frequency and duration of rainfall and snowfall. For instance, with every additional °C of global warming, the atmosphere is expected to hold seven per cent more moisture (IPCC, 2022a). An atmosphere with more water vapour can lead to more heavy rainfall events, potentially leading to flooding. At the same time, as the water cycle is altered, many places are suffering desperately from water supply shortages. In addition, the quality of the water can be affected by changed water quantity. When there is less water than usual, the water risks becoming more polluted as a small quantity of water dilutes and decomposes pollutants more slowly. When there is excess water in the river and it results in flooding, small pockets of contamination are easily dispersed by flood water to large areas.

Both water abundance and water scarcity can threaten human security. In the past two decades, water-related natural hazards have increased rapidly. Since the 1940s, 44 per cent of all the natural hazards have been related to floods. The number of floods has increased 134 per cent since 2000, compared to the previous two decades (WMO, 2021a). Floods can cause severe damage to human lives and the economy. Similarly, the scarcity of water can also cause damages, deaths and economic losses. Water scarcity has seriously aggravated globally and regionally as a consequence of climate change.

Currently, for at least some parts of the year, more than half of the world's population suffers from severe water scarcity (IPCC, 2022a). Reduced water availability can have large effects on local environmental systems, which in turn can have unprecedented social, economic, political and cultural impacts. For instance, drastic changes have been observed in soil moisture globally, which indicates that the water cycle has been altered, potentially triggering irreversible change. Consequently, in combination with deforestation, vital forests and rainforests such as the Amazon risk drying out. The Amazon is often referred to as the 'lungs of the earth' as it absorbs large amounts of greenhouse gases and releases oxygen. The potential environmental devastation will have serious effects on human lives and livelihoods. Human health and socio-economic development are already impacted worldwide as rivers, lakes and aquifers have been observed to have alarmingly low levels of water. Moreover, deteriorating water quality has become a serious problem globally, and climate change is making this situation worse. The quality of water is adversely affected by increased water temperature, extreme storm events and water shortages. On top of that, polluted water releases more greenhouse gas emissions compared to clean water.

The origins of the water crisis are not limited to climate factors alone. It has become an issue of concern for decades due to unequal distribution, unsustainable development and population growth. However, climate change is seriously aggravating problems related to water globally and regionally. Despite more than three decades of alarm bell ringing and big talk in global summits, the global water crisis has continued to become a massive challenge to humankind's survival and well-being. The UN 2023 Water Conference, which took place in New York on 22-24 March 2023, led to nearly 700 commitments from various agencies but none of them are binding to ensure universal access to clean water and sanitation. The water crisis is becoming more acute but the world is yet to agree on a common agenda to deal with the growing challenges.

WATER SCARCITY AFFECTING AGRICULTURE AND ECONOMIC DEVELOPMENT

Water scarcity is a growing threat to human populations and the environment. Population growth, industrialisation and urbanisation have increased the water demand leading to less water available to meet the growing needs. During the past 100 years, the demand for water has increased by 600 per cent to meet the growing demand for food and economic development (Wada et al., 2016b). By 2050, the demand for freshwater is expected to exceed population and economic growth. At the same time, the quality and amount of water resources available are expected to decrease. In fact, based on current observations of streamflow and evaporation, there is a prominent risk that the global water crisis is likely to get even worse than previously predicted, particularly in Africa, Australia and North America (Zhang, et al., 2023b). Climate change can significantly aggravate already existing problems of water shortages. In turn, it can cause serious consequences on human security and the environment.

Inevitably, water is key to socio-economic development. Increased water stress can therefore result in losses in income, health and agricultural production. The lack of rain and increasingly frequent heatwaves have alarmingly reduced water flows in many major rivers worldwide, from North America to Europe, from the Middle East to East Asia. This significantly affects food security, economic development and energy production in many parts of the world. During the summer of 2022, several important rivers in Europe faced severe water scarcity. As per the European Drought Observatory, nearly half of Europe was under a drought warning, which probably never has happened before in the last 500 years (Toreti et al., 2022). The low levels of water had substantial negative effects all over the continent. In France, farmers experienced a reduction of 12–15 per cent in their yields of main crops. Additionally, southwestern France experienced a reduction in their energy production in the nuclear power plants because the water from the river was too warm to cool down the facility. Italy and Portugal faced reduced energy production from hydropower plants due to low levels of water. Moreover, significantly low levels of water in the Rhine River caused shipping delays and thereby decline in economic prosperity (Toreti et al., 2022). The drying up of rivers and lakes has not only adversely affected irrigation, navigation and industrial production but has also led to a rising level of pollution as less water in the system fails to dilute even relatively common pollutants. In turn, water pollution in combination with water scarcity can double the risk of economic losses and can decrease the resilience of trade networks, as was the case in China in 2017 resulting from record-breaking drought in the northern region (Li et al., 2022).

Water scarcity can contribute to creating tensions and intensifying already existing disputes between water-sharing countries and entities. One of the serious effects is its consequences on the economy. The World Bank predicts that water deficits can limit countries' economic growth by up to six per cent of their GDP by 2050 (World Bank, 2022b). Low- and middle-income countries are expected to experience the highest GDP losses as a consequence of severe water stress (IPCC, 2022a). Mainly, because the economy in these countries is often more dependent on agricultural production, which is directly affected if the water availability is reduced (Swain, 2013). For instance, in the most water-scarce region in the world, the Middle East and North Africa region, agricultural production is expected to decrease by 25 per cent by 2080 due to precipitation deficits and increased temperatures (Swain and Jägerskog, 2016). According to NASA, the region has experienced drought continuously since 1998, accounting for the most severe dry spell in 900 years (Cook et al., 2016). UNESCO (2018) estimates that by 2050 an additional 2.7 to 3.2 billion people in Asia, Africa and the Middle East are expected to be living under severe water stress.

Urban areas are also highly at risk of increased water scarcity. In many parts of the world, urban areas are growing rapidly. By 2050, 68 per cent of the world's population is expected to live in urban areas. This could lead to an 80 per cent increase in water demand (UN, 2019). Consequently, critical services like water supply and sanitation will struggle to sustain the fast expansion of urban populations. Already now, many urban areas in arid regions in Africa, Asia and Latin America face threateningly low water supplies, which significantly slow down urban economic growth. Years of drought can slow down urban economic growth by 12 per cent, thereby even risking to revert development progress (Zaveri et al., 2021). In some cases, water has to be diverted from distant rivers and lakes in rural areas to supply the water demand in cities, which can create political tension between rural and urban populations (Swain, 2013). In other words, water deficits can have huge societal consequences which in turn can cause serious security implications. For instance, increasing competition for scarce resources can contribute to sparking societal unrest.

Water deficits can also have serious implications for food security which can bring turmoil to societies worldwide. The agricultural sector uses around 70 per cent of all freshwater globally and is highly vulnerable towards climate change with altered precipitation patterns and warming temperatures. Globally, around 60 per cent of all agricultural production is rain-fed, whereas 40 per cent is irrigated (FAO, 2021b). Climate change threatens one-third of global food production by the end of this century, following the current path of greenhouse gas emissions. Rapid and drastic reduction of greenhouse gas emissions would considerably reduce this risk. At the same time, given the large share of water used by the agricultural sector, increasing agricultural production is also a significant contributor to increased water stress. Particularly, because the water used in agriculture, mainly due to evaporation, does often not return to the water system where it will be available for other consumers to use, which instead can be the case for water used for other purposes (Swain, 2013).

When the precipitation is becoming less reliable, groundwater is becoming increasingly important for the agricultural sector. Groundwater accounts for the largest share of freshwater worldwide. Nearly 40 per cent of the world's groundwater is used for irrigation (UNESCO, 2022). However, groundwater is over-extracted in many places in the world. Consequently, it can lead to salinisation and exhaustion which aggravates already existing water stress (Ward and Ruckstuhl, 2017). On top of that, climate change directly affects the availability and quality of groundwater through changed precipitation patterns and weather extremes. The rising sea-level is also likely to lead to increased saltwater intrusion into coastal aquifers, which risks contaminating freshwater sources.

Growing water demand and an increasingly warming planet have not only resulted in the decline of water flow in river systems and aquifers, but several major freshwater lakes are also drying up. In 2022, water levels of the Poyang, the largest freshwater lake in China, dropped to a historic low due to an intense drought period. Climate change and unsustainable water use are drying up lakes in the Middle East. The water bodies of the Aral Sea in Central Asia, Lake Poopo in Bolivia, Lake Chad in Central Africa and Owens Lake in California have been continuously shrinking, creating a serious ecological crisis in and around these lakes and bringing more challenges to food production, trade and transportation.

THREAT TO HUMAN SECURITY: FOOD SUPPLY, HEALTH AND MIGRATION

Water's value is not only to be seen in economic terms. Clean and accessible water is an essential condition for life, and it also plays a fundamental role in human security. It is seriously worrying that the world still needs to be reminded about water's importance. Approximately one in three people in the world, two billion are living without clean drinking water; 3.6 billion people lack safely managed sanitation services; and 2.3 billion lack basic handwashing facilities (WHO and UNICEF, 2021). Several health issues, such as cholera, diarrhoea, polio, hepatitis A and typhoid are connected to poor sanitation and contaminated water. Despite being largely preventable, more than 800,000 people die every year from diarrhoeal diseases due to unsafe drinking water, sanitation and hand hygiene (WHO, 2022). Comparatively, about 237,000 people were killed in armed conflicts globally in 2022 (UCDP, 2023b).

In connection, human security is also threatened by the enormous challenge of ensuring food security for the world's growing population. During the last 50 years, the world's population has doubled from almost four billion in the mid-70s to almost eight billion in 2023. The amount of people facing hunger has increased by a total of almost 250 million people since 2019, being as many as 924 million in 2022 (WMO, 2023a). The gender gap of severely food insecure has also widened, with four per cent more women than men suffering from food deficiencies in 2021. Indeed, the outbreak of COVID-19 and armed conflicts also play their parts in the reduction of food security. Nevertheless, projections estimate that around 600 million people will be chronically undernourished in 2030, which is far away from the Sustainable Development Goal (SDG) to end hunger and ensure food access for all by then (FAO, 2023). The issue of malnutrition can be expected to worsen due to climate change as increased greenhouse gas emissions pose the threat of lowering the nutrition of protein, zinc and iron in crops (IPCC, 2019c). Meanwhile, in 2021, 3.1 billion people couldn't afford a healthy diet and almost 800 million were undernourished (FAO, 2022; WMO, 2023a). Rising food prices are a significant factor impairing the strive for food security, where low-income consumers and countries importing large shares of their food are particularly at risk. For instance, Arab countries in the Middle East and North Africa region import around 50 per cent of all calories they consume. Therefore, the region is significantly sensitive to large fluctuations in cereal prices (Swain and Jägerskog, 2016). Climate change, in combination with other non-climatic factors such as conflicts and pandemics, can create fluctuations in food prices which in turn can trigger new geopolitical tensions. Extreme weather events can destroy yields which disrupt the international trade of agricultural commodities and increase food prices. In combination with conflicts around the globe, such as in Ukraine, the global market for agricultural commodities is disrupted. In turn, it can exacerbate food insecurities worldwide. For instance, in 2022, the Horn of Africa experienced its worst drought in 40 years (Desmidt et al., 2021). The decline in yields due to the drought impeded smallholders' ability to hold down the local food prices when they were rising as a consequence of the war in Ukraine, leading to an even sharper increase in the food prices. This kind of situation can trigger geopolitical tensions particularly when drought occurs in several countries simultaneously and countries may impose export restrictions to protect the availability of food in one country, but diminish the access to food in another country (Anisimov and Magnan, 2023). Simultaneously as more than 10 per cent of the world's population is facing hunger and almost 40 per cent cannot afford nutritious food, insufficient and superficial measures are made worldwide to shift to less water-intense diets. In general, living standards are increasing and diets are shifting towards eating more and being increasingly dependent on animal-based products which are coupled with intense water consumption and higher greenhouse gas emissions (IPCC, 2019c).

Food and water insecurities can increase stress on poor and vulnerable communities which can force them to migrate. Forcibly displaced migrants are particularly at risk of food and water insecurity, regardless of being displaced by natural hazards or armed conflicts (Swain and Jägerskog, 2016). Many forcibly displaced people live in informal settlements or remote refugee camps in water-scarce areas. For instance, clean, accessible and affordable drinking water has been identified as one of the main priorities among forcibly displaced people in Arab countries. Urgent measures are needed to improve the water situation for forcibly displaced living in vulnerable settlements. The World Bank reports that 18 million forcibly displaced in Yemen don't have access to clean and safe drinking water, 25 per cent of Syrian households in informal settlements in Lebanon only have access to highly contaminated drinking water and water is regarded as unaffordable in 50 per cent of Libya's municipalities hosting forcibly displaced. These challenges were especially pressing during the outburst of COVID-19 which required increased water usage to ensure adequate handwashing and hygiene (Borgomeo et al., 2021). In addition, severe weather extremes such as flooding and heavy storms can lead to secondary displacements for people living in fragile refugee camps and relief centres. For instance, in Syria, severe winter storms led to more than 5,000 secondary displacements in 2022 (WMO, 2023a). In other words, water deficits and contamination can put additional burdens on already impoverished populations.

Several actions can be taken to build resilience to better cope with increasing challenges related to food and water insecurities due to climate change. These include policy measures addressing institutional and economic barriers to food security adaptations such as diversifying food systems, adapting to less water-intense diets, adopting a broader range of technological innovations in the food system and diversifying food production (IPCC, 2019c; Swain and Jägerskog, 2016; Ward and Ruckstuhl, 2017). The adaptation measures indicate that efforts can be made at various scales to ease the negative effects of climate change. For instance, policies can facilitate food security adaptations on an institutional level and strategies to reduce water waste can be implemented on a local level. Therefore, policymakers must focus on various adaptation measures. If not, there is a high risk that it will become even more difficult to deal with challenges connected with the changing climate.

DESTRUCTION OF HYDRO-PROJECTS, HIGH FLOODS AND GLACIER LAKE OUTBURSTS

Water scarcity and floods are the most hazardous water-related events for human populations. The World Bank reports that around 20 per cent of the world population is directly exposed to significant risks of extreme flood events. The majority live in South and East Asia, mainly in China and India. In total, almost 90 per cent of flood-affected populations live in low- and middle-income countries. Some of the poorest people exposed to risks of extreme flood events are mainly located in sub-Saharan Africa and South Asia. While populations in high-income countries also are exposed to risks of flood events, they are more likely to invest in mitigation, protection and recovery measures to ease the effects of extreme flood events. The most vulnerable areas are therefore where exposure to extreme flood events and poverty coincide. Floods also contribute to economic deficits, 31 per cent of all economic losses due to natural hazards between 1970 and 2019 were related to flood events (WMO, 2021b).

Given that climate change alters precipitation patterns, it increases the risks of extreme flood events. The melting of snow and ice at faster rates in glacier areas can add to the intensity of the floods. Along with melting snow water, the runoff from the glacial lakes can also contribute to extreme flood conditions. During the past two decades, the global glacier mass has decreased by more than 0.5 metre water equivalent per year (IPCC, 2022a). The glacier loss during 2021 and 2022 of the largest 40 glaciers was 1.18 metres water equivalent, which is a much larger loss reduction than the average during the past decade (WMO, 2023a). In the long run, however, one-third of the largest 56 glacierised catchments are estimated to experience reduced runoff by over 10 per cent by 2100, mainly in central Asia and the Andes (IPCC, 2022a). Currently, Asia and the Andes are the regions most at risk of floods as a consequence of glacial lake outburst floods (GLOFs) (Wang et al., 2020). A glacial lake gets formed as the ice in the glacier melts and the meltwater gets stored due to a wall of unstable debris being placed in front of the glacier. When these walls rupture, they release massive volumes of water downstream, leading to flash floods that can wreak havoc on communities and infrastructure. GLOFs are not a new phenomenon, but they have become increasingly common and devastating due to the changing climate. Over the past 30 years, the volume of glacial lakes globally has increased by 50 per cent due to the melting of glaciers caused by global warming. In turn, the frequency of devastating GLOFs is expected to rise. Around 15 million people worldwide are exposed to GLOF risks. Even more alarmingly, around half of this threatened population resides in just four nations: China, India, Pakistan and Peru (Taylor et al., 2023). It is important to consider both physical factors, such as the conditions of glacial lakes, and societal factors, including exposure and vulnerability when assessing GLOF risks. For instance, GLOFs are also occurring in European countries and Alaska. In Alaska, GLOFs have occurred 1,150 times during 35 years, which on average is around 33 events per year. However, these areas are often remotely located and therefore don't pose as severe risks for local populations and societies (Rick et al., 2023).

GLOFs in more populated areas without measures to mitigate the risks can cause detrimental damage to human lives and infrastructure and dramatically change the landscape. In 2020, a far-reaching GLOF triggered by heavy rainfall and/or melting process in the Peruvian Andes caused a significant number of fatalities and missing people in the valley downstream from the glacier lake (Vilca et al., 2021). Except for Peru, the most GLOF-affected countries are located in the Himalayas. The snow melting in the Himalayan Mountains (including Karakoram and Hindukush) affects all countries in the region whose river water systems are connected to the runoff. The countries are increasingly facing severe flooding in the monsoon months, bursting of glacial lakes in the spring season and drying up of rivers in the summer. The Himalayan Mountains have nearly 55,000 glaciers, and their snowmelt in the dry seasons feeds the 10 biggest rivers in Asia – from Indus to Ganges-Brahmaputra to the Mekong to the Yangtze. Most of the rivers are shared between two or more countries. The rapid glacier melting will most likely have significant socio-economic and geopolitical effects in the region.

India is one of the countries most affected by GLOFs. In 2021, flash floods triggered by GLOF in Uttarakhand in the Himalayas killed 80 people and left 124 missing. In this case, heavy snowfall precipitation in combination with a rapid glacier melts as a consequence of a sudden temperature change a few days earlier caused the flash flood. In the area, various hydro-projects were under construction. As a consequence of the flash flood, two hydropower projects were completely destroyed. Similarly, in October 2023, a GLOF in North Sikkim led to the collapse of a 1,200-megawatt hydropower dam and claimed over 40 lives. The flash flood directly affected almost 90,000 people, killed over 600 animals and damaged almost 2,000 houses and 16 roads (Sphere India, 2023). The increasing number of GLOFs in the Himalayan region not only leads to repeated humanitarian disasters but also adds further complexity to water resource management in the countries in the region. These events introduce sudden, unpredictable occurrences with far-reaching and devastating consequences for communities, infrastructure and ecosystems.

To reduce the risks, countries in the region need to develop early-warning systems, infrastructure resilience measures and cooperative disaster preparedness. Since many glaciers and glacial lakes in the Himalayas span multiple countries, cooperation and agreements between nations are necessary to address the shared risks and responsibilities associated with GLOFs. Given the increase in hydropower projects, infrastructure and population in affected Himalayan areas, more attention should be given to improving projection systems that account for extreme runoffs during lake outbursts (Veh et al., 2019). Despite being highly difficult and expensive to control and prevent flash floods triggered by glacial lake outbursts, it is less expensive than the economic losses caused by the GLOFs (Wang et al., 2020). To address the rapidly emerging GLOF crisis, the Himalayan countries also must prioritise climate action at the regional level and invest in climate-resilient infrastructure. Moreover, implementing early-warning systems can ease the effects of GLOFs and floods by allowing people to be aware of hazardous weather and prepare for its effects. Worldwide, only one-third of the population has access to early-warning systems, and it is mainly in the least developed countries and small island developing states where these systems are missing (WMO, 2023a).

PROBLEMS IN PLANNING WATER DEVELOPMENT PROJECTS

The increasing uncertainty in the water cycle and its effects on human populations and ecosystems instigate challenges for planning new water development projects. It is especially challenging for large infrastructure projects with long-term implications. As mentioned in the last section, unexpected flash floods can cause severe damage to water infrastructure projects. At the same time, water scarcity can also be a major challenge that in combination with future uncertainties can pose problems to the long-term planning of water development projects. Strategies to make water supply more reliable, such as large-scale investments to develop water infrastructure, water regulation and ensuring sufficient water for ecosystems can be subject to poor investment returns if the water availability fluctuates unexpectedly. Ultimately, the projects may need to be rebuilt or retrofitted to sustain the water supply given the new uncertain climatic conditions in the area. These kinds of measures can be highly costly. Particularly, it can put economically developing countries in a vulnerable position as they may require financial and technical assistance to even start undertaking large water projects.

Another major challenge related to water development projects is water's connection to energy, particularly hydropower. While energy transition has become the mantra of the world to mitigate climate change, hydropower has become an important component in many countries' strategies to move away from fossil fuel-based energy production. No doubt that the ambition of achieving net-zero emissions by 2050 can only be achieved by expanding renewable energy production. Hydropower constitutes the largest source of renewable electricity as it contributes nearly 60 per cent. Hydropower was accountable for 17 per cent of the total global power produced in 2020. In the last two decades, hydropower generation worldwide has increased by almost 70 per cent, and it is estimated to increase another 50 per cent in the next two decades. It is mainly prioritised by economically developing countries for the sake of national energy security. At present, there are 1,000 dams under construction. The massive dam-building is ongoing in China, India, Turkey and Ethiopia. Brazil, Nepal, Laos and DRC have a long list of planned dam projects (IEA, 2021c).

Despite the common impression of hydropower being 'green' energy, it emits greenhouse gases that often are overlooked in the climate change debate. The connected large dams and reservoirs inundate large areas causing vegetation fermentation and removing massive tree cover. Research finds that the rotten vegetation behind dams emits nearly one billion tonnes of greenhouse gases every year, around 1.3 per cent of total annual human-caused global emissions. Additionally, nearly 80 per cent of greenhouse gases emitted by dam reservoirs are methane, a gas which is a highly accelerating agent in causing global warming (Deemer et al., 2016). The contribution of greenhouse gases from dams varies from region to region. In temperate and dry climate zones, some of the dam reservoirs might work as carbon sinks, absorbing more emissions than what is released. Conversely, in tropical countries, where most dams are being built, the carbon footprints are equal to or even greater than fossil fuels (Räsänen et al., 2018). Therefore, countries that currently are at the forefront of this trend should be extra careful before taking up dam-building as their strategy to mitigate climate change because the benefits may not be as expected. In some cases, particularly in the short term, it can even accelerate global warming.

Moreover, hydropower comes with a high social, economic and environmental cost. For instance, it is highly costly to build dams which currently make it unattainable for some countries with lower economic capacities. Dam construction can also cause major conflicts between groups and nations. Globally, the hydropower industry has displaced around 80 million people due to dam projects, primarily indigenous and frontline communities living in forests and rural areas (Swain, 2010; Walicki et al., 2017). In addition, dams also destroy the free-flowing river ecosystem, block fish migration, erode downstream river beds and adversely affect the river's aquatic and riparian life. In other words, the challenges in planning new water development projects must be dealt with by bringing water professionals and political leaders to work together and create a synergetic water governance system aiming to better cope with climate change-induced uncertainties.

CHALLENGES FOR NEW WATER-SHARING AGREEMENTS AND THREATS TO EXISTING ONES

To complicate water management further, many of the world's rivers, lakes and aquifers are crossing borders and shared between two or more countries. Globally, around 310 rivers and lakes are crossing national borders, as is illustrated in Map 5.1. Moreover, more than 500 aquifers cross national borders (UNESCO, 2021). These river basins are home to 52 per cent of the world's population. Shared river management has become the priority of nation-states and multilateral agencies. Since 805 AD, more than 3,600 treaties have been made between countries to manage the water of these rivers and the last 200 years have seen the signing of 600 of them (Eckstein, 2017). Agreements to share cross-border aquifer systems are much less common. Globally, only six legally binding agreements and five informal sharing arrangements exist for this kind of water source.

Formal and informal norms and institutions have been developed among the basin countries from Colorado to the Rhine, from the Danube to the Euphrates-Tigris to the Aral Sea over the water-sharing. However, these existing water-sharing mechanisms or basin-based water management institutions are proving inadequate to guide the path of cooperation over freshwater given that climate change has brought unprecedented reduction to its accessibility.

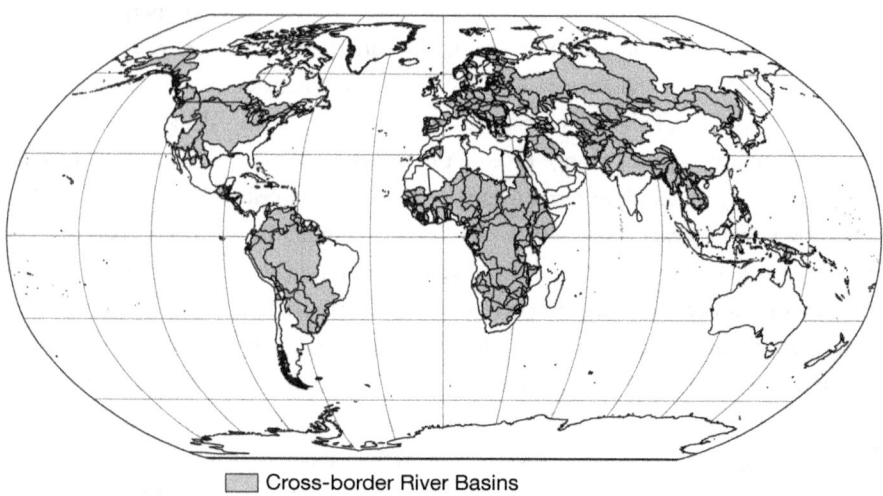

Map 5.1 River Basins Crossing National Borders

Note: The map shows the 313 identified river basins crossing national borders globally.
Source: Product of the Transboundary Freshwater Diplomacy Database, College of Earth, Ocean, and Atmospheric Sciences, Oregon State University. Additional information about the TFDD can be found at: https://transboundarywaters.oregonstate.edu

Still, despite the available knowledge, legal regimes and institutions focusing on cross-border rivers, many challenges persist. While regional and local politics complicate the policies towards efficient management of shared water resources, the threat of global climate change is increasingly undermining the existing sharing arrangements.

According to the World Meteorological Organization, more than 20 per cent of the world's river basins experienced either rapid increases or reductions in their surface water area in 2020 (WMO, 2021a). As previously discussed, rapid changes in water flow in the shared river systems contribute to increased risks of natural hazards like floods and drought, breaking down existing water-sharing arrangements and enhancing the dangers of escalation of water-sharing conflicts between countries. Adding further to the problem of water scarcity, the global climate change-induced uncertainty over water availability has started undermining the ongoing water-sharing arrangements on cross-border rivers. Basically, challenges to sharing water often increase during summer months with less water available for cities and farms for irrigation while the demand is high, and in arid and semi-arid areas the water issues are likely to be treated as security challenges (Swain, 2013). Many of the existing water-sharing agreements don't have the capacity to address the emerging challenges connected to climate change. For instance, the 30-year agreement on the Ganges water-sharing between India and Bangladesh will come to an end in 2026, and the climate change-induced uncertainty about the river flow raises serious apprehension about the possibility of the agreement's renewal. While the old water agreements on rivers and lakes are under severe pressure, it is becoming even more challenging to sign new agreements on water-sharing. This poses a serious threat as changes in the environment can aggravate political tensions, especially if no agreement is in place. In the case of the Nile, the lack of consensus on how to share the water during the prolonged drought period has become the main hindrance to reaching an agreement between Egypt, Ethiopia and Sudan.

Lastly, despite its importance, groundwater rarely reaches the public discourse or political agenda at the national or global level. The main reason is that the groundwater, unlike the water in the rivers and lakes, is not visible, so being out of sight has helped it be out of the fight. The quantity, quality and flow of groundwater are indiscernible to human eyes, prohibiting groundwater from getting the policy priority for its protection. Much is still unknown about the hydrological and geological characteristics of many aquifers worldwide. Still, groundwater is the most extracted natural resource on earth. One thousand cubic kilometres of groundwater are pumped out daily, and the amount is so high that it also contributes to sea-level rise (Wada et al., 2016a). Pulling water from the ground is commonly preferred over surface water because of its reliable supply and quality. But, when the water table drops below 15-20 meters, in most cases, it becomes economically impractical and water quality drops. Therefore, it is highly worrisome that such a small number of agreements regarding sharing aquifers exist while the water scarcity situation becoming further precarious, particularly due to climate change.

DEBATES AROUND THE POTENTIAL FOR CREATING 'WATER WARS'

The water crisis has become such a magnitude that it is growing into a common global concern. The situation is bound to get exponentially worse. Besides the apparent poverty creation and human development impediments held up by water insecurity, it can also be connected with increased political tensions. Ultimately, disputes over shared water can be triggered into armed conflicts, between or within countries. As climate change brings changes to water supply and demand patterns, the existing arrangements of sharing water resources between and within countries in the arid and semi-arid regions are likely to be more and more conflictual. There is no doubt that the projected impacts of global climate change on freshwater may be huge and dramatic, but they may not be as intense or follow similar periodic patterns in each region. While climate change may not cause a water war by itself, it aggravates these kinds of threats in different parts of the world. Water-scarce regions like Asia, Africa and the Middle East are particularly at risk.

Though countries have signed hundreds of agreements and no war over water has been recorded; there has often been apprehension and speculation of wars erupting over river water. Disputes over rivers often occur due to quantity, quality or control. Of these, a reduced amount of water is more difficult to address because water is not easily replaced (Swain, 2013). Although the world is yet to witness a water war, that possibility can't be written off as climate change has created huge variations and unprecedented uncertainty in water availability and demand in shared water systems. Nevertheless, armed conflict is only one of the various potential forms of action related to disputes over shared water. Debates, disagreements, demonstrations or measures to improve water management can also be outcomes of political water tensions (Swain and Jägerskog, 2016). Though there has been no war fought yet exclusively on water issues, many wars have taken place in the past, particularly in South Asia and the Middle East, whose origins had strong water footprints.

Models used to understand the relationship between water scarcity and conflict have a complex and circuitous anatomy. Identifying causal pathways from water scarcity, to the formation of actors, issues and actions that may escalate hostility into violence is cumbersome and rarely direct. Some studies have found causal links in a few cases, but this isn't enough to enable firm prediction. Instead, climate-induced water insecurity is more likely to act as a catalyst or a 'threat multiplier' that aggravates societal insecurities that may result in conflict (Raleigh et al., 2008; Swain and Öjendal, 2018). Decreasing water availability can weaken societies' mitigation and adaptation capacities in countries that already are experiencing economic and political challenges. Societies become highly vulnerable when they suffer from poor central leadership, weak institutions and polarised social identities. Governance, elites, institutions and social identities can make a difference between adaptation and confrontation (Swain et al., 2011).

However, conflicts between countries sharing water are not uncommon. For instance, the number of conflicts between upstream and downstream countries of the rivers originating from the Himalayas has increased as less than usual water flow in the summer season and more than the usual water flow in the monsoon period. When a downstream country gets affected by the flood, it blames the upstream country for releasing more water than usual. The dwindling water flow in the river system during the summer season has been a regular point of contention between upstream and downstream too. The fast and early melting up of glaciers brings less water to the rivers in summer and that can potentially lead to more disputes between the basin countries. The widening gap between the demand and supply of water is driving many disputes as it leads to competition, contests and tensions. Political tensions between riparian states further aggregate water-related disputes, such as between Pakistan and India (Jarvis, 2014). Water disputes can also derive from asymmetrical power relations between countries sharing the water, where unequal water distribution or agreements can spur conflicts (Ward and Ruckstuhl, 2017).

While the main focus previously and continuously is on shared rivers, the management and sharing of the groundwater, which is the primary source of freshwater supply, has not become a priority. As there is growing insecurity over water availability due to climate change, the receding groundwater table can cause more intense conflicts between aquifer-sharing countries. The risk of 'groundwater wars' can't be overlooked if the world fails to give adequate importance to groundwater management as it deserves. There is no doubt that climate change poses extreme challenges to water-sharing, and it has all the potential to create political instability and violent conflicts. Climate change requires countries to have more flexible, hands-on and politically smart water resources management. But that is not happening. Instead, many rivers have already become the source of bitter conflicts, particularly in Africa, the Middle East and South Asia. The increasing threat of water conflicts will be seriously damaging to the strive for efficient water management which is needed to mitigate and adapt to climate change (Swain and Jägerskog, 2016).

DEBATES AROUND WATER ACCESS AS A SOURCE OF COOPERATION

Changes in water availability often lead to increased water-related interactions between concerned actors. While political instability and ultimately water wars can be one potential outcome, another one is the establishment of cooperation between water-sharing actors. Basically, they are two ends of a spectrum with water-related actions, where it is also possible that disputes and cooperation can coexist in the same area (Swain and Jägerskog, 2016). The dominant discourse in research favours the idea that water deficits more likely lead to cooperation rather than conflict. Declining water access can influence the actors' understanding of the necessity to cooperate over the resource, rather than taking the risk of a serious conflict and bringing further insecurity over the future availability of a precious resource. Moreover, with the large share of freshwater being stored in cross-border sources worldwide coupled with the growing water insecurities posed by climate change, nation-states are becoming increasingly aware that they will not be able to solve their water issues alone.

Given the vast number of water-sharing agreements signed and implemented worldwide, it might not come as a surprise that cooperation over cross-border waters has been the dominating pathway for countries sharing water sources (Bernauer and Böhmelt, 2020; Wolf, 1998). Historically, access to water has played a major role in the creation of societies. Water, in general, and rivers, in particular, have been the source of nation-building in the past. Dynamic cultures and great civilisations have grown across the river resources. Water has brought many countries and cultures together. For instance, cooperation over the Rhine River has most likely laid the foundation for establishing the present European Union. Mekong, Ganges and Zambezi rivers have also set the stage for other forms of cooperation among their basin countries. The positive contribution of cooperative arrangement over the Colorado, Columbia and Limpopo Rivers to their basin states' bilateral relations is significant. Even India and Pakistan are cooperating over the sharing of the Indus water since 1960 despite fighting at least three wars during this period on other issues.

However, solid and sustainable water cooperation can be expected to be more likely between countries sharing similar democratic and legal systems, particularly because higher levels of trust can be expected between these countries (Zawahri and Mitchell, 2011). Nevertheless, water cooperation can be formed in various ways, formal or informal. Some agreements are bilateral, others are basin-wide. Globally, the 1997 UN Watercourses Convention provides a general guideline one sharing and managing cross-border rivers, though only 33 countries are party to it. In other words, the world is far from agreeing to a 'formula' on how to share its 310 international rivers, covering 150 countries and half of Earth's land surface.

Water cooperation does not only imply that no conflict is present. Instead, the mutual will to establish cooperation based on communicative and peaceful means should drive the process. This entails mutual problem-solving mechanisms and negotiation commitments (Swain and Jägerskog, 2016). One of the most structured ways to cooperate over shared water sources is through river basin organizations (RBOs) (Mitchell and Zawahri, 2015). RBOs have many times been essential to address and mitigate water-related problems in shared waters. However, their contribution to effectively and sustainably governing river basins has not always been sufficient (Swain, 2012a). The mere existence of cooperative agreements or arrangements is not enough to ensure the efficient water management needed to cope with the emerging global water crisis (De Stefano et al., 2012; Mitchell and Zawahri, 2015). For instance, the agreements or arrangements can be limited to only cover a few issues or may uphold prevailing power asymmetry between riparian states. Therefore, it may be insufficient to tackle the global water crisis due to climate change with all its complex components discussed in this chapter. However, a comprehensive approach is needed to create long-lasting and sustainable cooperation. This requires treating the river system as a single unit, recognising local social and cultural contexts related to water use, involving both state and non-state actors, as well as creating clear water-sharing rules and establishing a network to exchange information (Swain, 2013).

While the water issue is becoming extremely serious particularly due to climate change, the world in general and several large countries in particular lack an effective legal and policy framework to address this challenge. Global leaders and negotiators must realise that the centrality of water for the survival of human beings and the planet can't change. The only thing that can change is how we use and manage it. There is no shortage of technology or knowledge about better management to overcome the growing water challenges worldwide. The only key missing link is the willingness of the countries to work together. They still believe that they can manage the water issue on their own by focusing solely on the demand and supply of their respective countries. Unfortunately, political leaders have found it convenient due to societal insecurity over water scarcity to do politics over water. The rise of populism worldwide has made it worse as water is being projected through the ethno-nationalist prism. It is becoming almost impossible to put the existing water management ideas into action.

The political leaders are not only at fault. Even the water professionals with their 'we know it all' mindset don't care to understand the politics and political hazards while framing the water management strategies. Given the threat of water scarcity, it is essential to take politics into account while crafting water management strategies. And, at the same time, political leaders need to learn to keep water out of their politics and delegate the water professionals the power to manage this highly precious and increasingly scarce resource. It is easy to foresee that climate change will force comprehensive adjustments in the ongoing water management mechanisms as they need to have the flexibility to adjust to the uncertainties. The emerging unprecedented situation due to changes in climatic patterns requires countries and regions to cooperate and act collectively. The world's water crisis is real and serious; climate change has made it exigent – there is no time for the blame game; it is high time for them to work in tandem. It is high time that the countries must work together in partnership to find ways and means to meet the challenges posed by the water crisis successfully.

SUMMING-UP

This chapter has brought to the forefront the multifaceted challenges posed by climate change on water security, illuminating a scenario where the escalating global water crisis not only intensifies poverty and hinders human development but also seeds political tensions that could escalate into armed conflicts. The situation is most dire in regions like Asia, Africa and the Middle East, where water scarcity is a persistent issue. Climate change, acting as a 'threat multiplier', exacerbates these challenges, transforming the way countries must approach water resource management.

The complex dynamics of shared water resources, often leading to conflicts in areas with unequal water distribution or power imbalances are common. Yet, amidst these challenges, there exists a beacon of hope – the potential for water scarcity to foster cooperation. Historical precedents, such as the collaborative management of the Rhine, Mekong and Ganges rivers, demonstrate the possibility of turning a crisis into an opportunity for unity and collective action. However, the effectiveness of such cooperation is contingent upon the political and legal harmony among the countries involved. The mixed success of RBOs underscores the necessity for a nuanced and inclusive approach to water management.

The current global frameworks, or the lack thereof, is not effective enough in addressing the water crisis exacerbated by climate change. Political leaders need to transcend the politicisation of water issues, while water professionals must be aware of the importance of considering political contexts in their strategies. Only a flexible, adaptive approach to water management can respond to the unpredictable shifts brought about by climate change.

In the face of climate-induced water crises, global cooperation is imperative due to the interconnected nature of our ecosystem and the universal challenges posed by climate change. Many of the world's critical water resources, such as rivers, lakes and aquifers, traverse national borders, making international collaboration essential to manage these shared resources sustainably and equitably. The far-reaching impacts of climate change, including altered weather patterns and extreme weather events, disregard national boundaries, necessitating a unified response. Tackling the water crisis demands collective action and shared knowledge, leveraging pooled resources, technological innovations and best practices in water conservation. Moreover, international cooperation plays a crucial role in preventing conflicts over scarce water resources, ensuring regional stability, and upholding human rights, particularly the fundamental right to clean water. Economic interdependence further underscores the need for joint efforts, as water scarcity can disrupt global trade and food markets. Shared risks such as droughts, floods and storms brought on by climate change call for cooperative adaptation strategies. Therefore, addressing climate-induced water crises effectively demands a collaborative approach that prioritises equitable access, conflict prevention and ecosystem protection on a global scale.

A collaborative approach, underpinned by comprehensive and sustainable water management practices, is essential. As the global community faces these unprecedented challenges, it becomes increasingly clear that no single nation can combat the impacts of climate change on water security alone. The path forward requires a concerted effort, transcending national boundaries and political interests, to ensure a water-secure future for all. The urgency of the situation cannot be overstated – it is a call to action for countries worldwide to unite in their efforts to address the profound implications of climate change on our most precious resource: water. In this context, the climate issue needs to be elevated as a national security concern so the leaders can be politically and administratively able to look beyond the short-term gains and negotiate on critical water-sharing issues.

6
CLIMATE MIGRATION

As the dawn of the 21st century unfolds, a silent but profound exodus is reshaping our world. The chapter 'Climate Migration' delves into the complex tapestry of human movement spurred by the relentless march of climate change. In the face of rising seas, scorching droughts and devastating natural disasters, a new type of migrants emerges – not by choice, but by the sheer force of survival. This chapter unravels the intricate interplay of environmental degradation, sociopolitical dynamics and the human spirit's resilience, painting a portrait of a world in transition and the extraordinary challenges we face in accommodating and understanding those displaced by the very earth they call home.

POSSIBILITY OF MASSIVE POPULATION DISPLACEMENT

The devastating effects of climate change can force people to migrate from their home communities if they are unable to sustain their livelihoods due to the changing environmental conditions. Already in the beginning of the 1990s, the Intergovernmental Panel on Climate Change (IPCC) warned that large-scale migration could be the most significant single impact of climate change on human populations (IPCC, 1992). Predictions are that the scale of this mass migration will substantially increase in the near future. Despite the early warnings from the IPCC, climate-induced migrants have until recently been viewed as a peripheral concern for the international community. However, the increasing impact of global warming and sea-level rise has brought climate-induced migration to the fore as a critical issue on the international political agenda. Mainly, because large-scale migration risks lead to tensions, conflicts and humanitarian crises if the emerging challenges are not handled properly. In particular, climate-induced migrants risk facing hardships to migrate or be accepted in a new community in the absence of any international legal protection for migrants in this category. Sea-level rise may be one of the most obvious climate-induced environmental changes that can force people to migrate. However, as climate change can exacerbate already existing environmental problems, the world must be prepared for other kinds of climate-induced displacements as well.

People migrating in response to changing environmental conditions is not a new phenomenon. In order to escape floods, fires and droughts, people have always migrated in search of better living conditions (Jägerskog and Swain, 2024). Deforestation and desertification have also had a significant impact on population migration in the past. Now, the consequences of climate change have raised the likelihood of mass migration on a scale the world has never witnessed before. A growing number of people already are moving away from their homes because life has become insupportable due to environmental changes (IDMC, 2023; WMO, 2023a). As climate change can exacerbate environmental changes and connected vulnerabilities, it is expected to force an increasing number of people to migrate. However, it is too simplistic to conclude that people will start moving solely because the environment is changing. Instead, various factors, such as social, political or economic factors can influence if and how someone chooses to or is forced to migrate.

The complexity of migration, especially when it is induced by climate change, makes it difficult to define climate migration. Environmental migration was systematically defined by the author of this book in the 1990s as migrants who are forced to move away from their homes as a result of the loss of their livelihood and/or living space due to environmental changes (natural as well as anthropogenic) and migrate (temporarily or permanently) to the nearest possible place (within or outside the state boundary) in search of their sustenance (Swain, 1996a).

Since then, various attempts to define environmental migration have been made. Mainly, it concerns changes in the environment that push people to migrate. Climate change has brought another nuance to the definition where climate migration is defined as: the movement of a person or groups who, predominantly for reasons of sudden or progressive change in the environment due to climate change, are obliged to leave their habitual place of residence, or choose to do so, either temporarily or permanently, within a State or across an international border (IOM, 2019:31).

In other words, climate migration, just like environmental migration can be short term or long term and can be within or between countries. The most important aspect of the term 'climate migration' is that the migrants are forced to leave their home communities because of climate-induced changes in the environment. However, as migration often consists of the influence of various factors, it can be complex to determine the role of climate change in the migration response. Therefore, it is difficult to argue that it is solely or mainly climate change that pushes people to migrate. Several push and pull factors can drive migration, which of many can be unpredictable and not easy to assume because of their complex interaction. Push factors can consist of a deteriorating environment, poor economic conditions or political instability. Pull factors are conditions that can make people attracted to migrate to a new location, such as increasing economic opportunities, better health care or improved living conditions (Swain, 2019). This means that changing environmental conditions due to climate change can be a strong push factor, which in complex interaction with other push and pull factors can lead to migration, forced or voluntary. Undoubtedly, the social, religious, ethnic, political, economic and demographic drivers play their parts. Migration becomes the outcome of a complex and multifaceted dynamic of interaction between different factors (Jägerskog and Swain, 2024).

The changing environmental conditions that can push people to migrate can be both sudden and slow. Sudden changes can be floods, storms and wildfires that instantly can destroy livelihoods, infrastructure and lives. These environmental hazards are predicted to become more frequent and intensified due to climate change. Slow changes can be droughts, sea-level rise, extreme heat or declining water availability. The slow changes can push migration through cumulative effects repeated over extended periods, such as declining agricultural yields (Clement et al., 2021; Zaveri et al., 2021). As climate change risks exacerbate the severity of environmental changes, already existing vulnerabilities risk increasing and thereby risk inciting migration rates to unprecedented levels, especially if insufficient measures are taken to improve the situation locally. When people's livelihoods are severely affected by environmental changes, and when there is a lack of substantial opportunities to provide for themselves in their home society, this kind of migration should essentially be considered forced displacement (IDMC, 2021a;b). Countries and societies cope with climate-induced challenges differently depending on their economic and institutional strength. Some countries and societies have more potential to plan and implement adaptation strategies to meet climate-induced challenges. For instance, investing in water supply and sanitation, while ensuring a flexible and active labour market can increase people's resilience (Jägerskog and Swain, 2024; Zaveri et al., 2021). The implemented strategies can either increase or decrease people's vulnerability, which in turn can impact how and if people migrate.

CLIMATE MIGRATION AND SECURITISATION OF NATIONAL BORDERS

Despite the widespread consensus that exacerbated climate change will induce large-scale migration, the number of people currently migrating or predictions of future migration numbers as a response to the changing climate is uncertain and debated. Estimating global migration rates is a difficult task. Mainly, predictions can be unreliable as countries don't share uniform definitions of migration, which complicates data collection and comparison. In many countries, available data on migration is often incomplete and unreliable. Additionally, given that climate change may exacerbate uncertainties, it complicates predictions further. Currently, the predictions of the number of climate-induced migrants by 2050 vary from 25 million to 1 billion. The common average is estimated to be 200 million migrants by 2050 (Swain, 2019). For instance, the World Bank estimates that around 216 million people across six regions in the world could move within their countries by 2050, a number that is predicted to increase significantly through the second half of the century. The estimate is divided by region as follows: 86 million in Sub-Saharan Africa, 49 million in East Asia and the Pacific, 40 million in South Asia, 19 million in North Africa, 17 million in Latin America and five million in Eastern Europe and Central Asia (Clement et al., 2021). By 2100, sea-level rise alone can displace more than 200 to 600 million people by 2100 (Kulp and Strauss, 2019). Another estimate is that around 410 million people will be highly vulnerable to the rising seas by the end of the century (Hooijer and Vernimmen, 2021). There is no agreed estimate on how many climate migrants will move beyond their national borders. Nevertheless, there is no doubt that climate change is already displacing many people and forcing them to move to other countries and regions in search of survival (Behrman and Kent, 2022).

The International Displacement Monitoring Centre (IDMC) reports that the number of internally displaced by natural hazards has surged in recent years. In 2022, 31.8 million people were reported as displaced by natural hazards such as storms, floods and droughts, as shown in Table 6.1. The reported displacements were 41 per cent higher than the annual average of the past ten years. Out of the 31.8 million people, 25 per cent were displaced due to the flooding in Pakistan, and at least 1.1 million due to the worst drought in 40 years in Somalia (IDMC, 2023). In Somalia, at least 600,000 people crossed the border between Ethiopia and Kenya (WMO, 2023a). In Pakistan, around eight million people were internally displaced as a consequence of the severe flooding. In the Indian state of Assam, at least 600,000 people were displaced due to flooding during the same year (WMO, 2023a). A severe flooding in Libya in 2023 displaced around 50,000 people and caused thousands of confirmed and unconfirmed deaths (IOM, 2023). In 2020, 2021 and 2022, the number of people displaced by natural hazards was accountable for more than 50 per cent of all displacement, including displacements by conflicts and violence (IDMC, 2022). In 2022, 387 natural hazard events were reported globally, affecting 185 million people. Asia is the continent most affected by natural hazards, where more than 35 per cent of all the natural hazards occurred in 2022 (CRED, 2023). Undoubtedly, natural hazards already force millions of people to migrate. Though the monitoring of internally displaced people is substantial, it is prominent that there are many unreported cases of both internal and international migration as a response to the environmental changes induced by climate change.

Table 6.1 Internal Displacements by Natural Hazards in 2022

Hazard	Number of Internal Displacements
Floods	19,219,000
Storms	9,980,000
Droughts	2,215,000
Wildfires	366,000
Landslides	53,000
Extreme Temperatures	12,000
Total	**31,845,000**

Note: The table shows the amount of displacements per natural hazard during 2022.
Source: Internal Displacement Monitoring Centre (IDMC).

Currently, a majority of reported climate-induced forced displacements are connected with sudden environmental changes, such as floods and storms. However, flooding does not always lead to long-term migration. For instance, flooding in Pakistan from 1991 to 2012 only had a marginal effect on long-term migration. Flooding during the period attracted a substantial amount of disaster relief. Heat stress, on the other hand, had a stronger effect on long-term migration, while also attracting less disaster relief. Heat stress and drought during the period seemed to have a greater effect on people's income and thereby had a stronger effect on migration as an outcome (Mueller et al., 2014). An increasing number of people are being reported to be displaced by slow environmental changes, such as droughts. For instance, in the 1990s, at least two million Bangladeshis migrated to India after being displaced by water insecurity (Swain, 1996b). However, the complexity of this kind of migration makes it more difficult to monitor compared to sudden environmental changes (IDMC, 2021a). For instance, in the world's most water-scarce region, the Middle East and North Africa (MENA) region, only Iraq has reported drought-induced displacements. However, the IDMC deems that evidence shows that more countries in the region have similar experiences (IDMC, 2021b). Therefore, a vast number of unreported cases of displacements by slow environmental changes can be expected globally.

Water insecurity, when there is less water available than demanded, can increase economic hardships for rural populations which in turn can induce migration processes. The economic impact of periods of droughts can make it less obvious to determine if migration is a reaction to the changed environmental conditions rather than the economic decline. However, when water insecurity is the main reason for economic decline, the environmental condition becomes a major push factor (Afifi and Jäger, 2011). The World Bank estimates that declining water access has contributed to a 10 per cent increase in global migration rates between 1970 and 2000. During this period, water scarcity is explained to be a five times higher driver of migration compared to water excess, such as floods. Water insecurity can be the result of water mismanagement, sudden population growth and/or climate change (Zaveri et al., 2021).

However, displacements by floods are often easier to monitor as the effects of the sudden environmental changes are more detectable.

Mainly, migration connected to declining water access is predominantly occurring from rural areas dependent on agriculture (Borgomeo et al., 2021; Zaveri et al., 2021). The Middle East and North Africa are particular hot spots where people are expected to migrate because of declining water availability. In particular, the most profound hotspots in North Africa are the northeastern coast of Algeria, western and southern Morocco and the western part of the Nile Delta in Egypt (Clement et al., 2021). Undoubtedly, though migration is a complex phenomenon, access to water can play a significant role, especially for people who are dependent on agriculture to sustain themselves (Jägerskog and Swain, 2016). People generally struggle to survive in a water-scarce situation for long periods before they migrate (Swain, 2019). The expected hotspots of host locations for migrants from water-scarce areas in North Africa are the urban centres of Cairo, Algiers, Tunis and Tripoli (Clement et al., 2021). However, it is also important to highlight that not everyone can migrate, even if the migration is forced. Migration can be costly. Water scarcity often has negative effects on people's incomes. Especially, for people working in agriculture where their yields may be destroyed when there is not enough water. Therefore, poorer people may be trapped in deteriorating environmental conditions if they can't afford to migrate. In fact, the World Bank reports that people in upper middle-income countries are four times more likely to migrate as a response to water scarcity compared to people in low or lower middle-income countries (Zaveri et al., 2021). In other words, while climate change risks increasing the number of people displaced by water insecurity, it also risks trapping people in water-scarce conditions.

Another change in the water conditions that can force people to migrate is sea-level rise. Sea-level rise is one of the most prominent examples of climate-induced migration. The rising seas threaten many low-lying coastal areas, including mega-cities such as Bangkok, Ho Chi Minh City and Manila. Low-lying small island states are also highly affected by sea-level rise. At least ten per cent of the global population lives in low-lying coastal areas less than five metres above the high tide line (Coalition for Urban Transitions, 2019). Sea-level rise is taking away the living space and source of livelihood of millions of people. The IPCC warns of the possibility of a sea-level rise of up to 1.1 metres by 2100 if emissions are not reduced drastically (IPCC, 2019d). A sea-level rise of more than one metre will not only inundate large low-lying areas in many countries but can potentially submerge many small island countries. Already in 1987, the then President of the Maldives, Maumoon Abdool Gayoom told the UN General Assembly that a sea-level rise of only one metre would be catastrophic for the nation, and risk submerging the entire country (UN General Assembly, 1987). Nearly four decades have passed, and the threat of several small island countries disappearing from the global map altogether looks more acute than ever before.

The rate of global sea-level rise has doubled in the last three decades, from a global mean average of 2.27 mm per year between 1993 and 2002 to 4.62 mm per year between 2013 and 2022 (WMO, 2023a). Currently, small island states in the Western Pacific are facing a sea-level rise of more than four times the global average. The effects already have severe effects on local livelihoods. For instance, people living on the atoll Carteret Islands of Papua New Guinea have been forced to be relocated to resettlement camps on Bougainville Island because of the rising seas. This displaced community is often referred to as the 'world's first climate refugees' (UNFCCC, n.d.). Similarly, inhabitants of the Gardi Sugdub Island in the Caribbean Sea are expected to become the first climate refugees in America. The island is located outside of Panama's northern coast and has been inhabited by the indigenous Guna people for at least 100 years since they resisted forced assimilation after Panama's independence. Now, the island is increasingly suffering from flooding as a result of the rising sea. As a consequence, the island is progressively becoming less habitable. The leaders of the island, together with the government in Panama are working on establishing a new village on the mainland, starting with the planned relocation during spring 2024 (Salido, 2024). Except for the immediate threat of rising sea-levels, small island states also face other climate-induced environmental challenges that in the long run can force people to migrate. For instance, the increasing seawater temperature and ocean CO_2 concentration adversely affect coral reef systems. Coral reefs play a big role in the well-being of small island countries by supplying sediments to island shores and restraining the impact of waves. They are also home to some of the most biodiverse and productive ecosystems on our planet. At least half a billion people worldwide are dependent on coral reefs for their food, income and protection (NOAA, 2019). Changing rainfall patterns, decreasing precipitation and increasing temperature have also presented critical challenges for the freshwater supply in these islands and to their food security. In other words, climate change poses many challenges to small island states and low-lying coastal areas. If adequate strategies to mitigate climate change are not enforced, there is a prominent risk of massive climate-induced population displacement.

Clearly, climate change and environmental changes have already forced people to leave their homes and move to other areas. Currently, it is the most vulnerable developing countries that are disproportionately affected by climate-induced migration. Many of the climate migrants become internally displaced people or end up in immigrant detention camps in neighbouring countries (Rosignoli, 2022). Mainly, restrictive migration receiving policies imposed by many countries have contributed to the substantial increase of internally displaced people forced to leave their homes because of environmental changes, as well as wars and armed conflicts (Swain, 2019). Nevertheless, a significant portion of climate migrants is crossing national borders. The increasing international climate migration phenomenon is a growing concern for the international community as it can create security concerns for nation-states (Jägerskog and Swain, 2016). Worldwide, large-scale cross-border migration can be perceived to pose challenges to the peace and security of many nations. The perceived security challenges related to increasing migration flow cause some receiving nations to face critical policy challenges where increasing attention is given to the securitisation of national borders. In particular, by framing climate migration as a threat to national security, many countries are focusing on restricting legal entry. For instance, the Global North has a growing focus on protecting its domestic and regional borders. While shifting focus to the securitisation of national borders, less attention is given to protecting climate migrants and their rights (Behrman and Kent, 2022). To ensure human security, however, the challenge of increasing numbers of forced migrants must be addressed by setting up a comprehensive agenda where more attention is given to preventing causes of displacements rather than the use of force to stop international migration. In fact, early and decisive action to strengthen climate mitigation and adaptation strategies can reduce the number of people migrating internally by 80 per cent by 2050 (Clement et al., 2021). At the same time, the international community must ensure that people who are forcibly displaced have access to their basic human rights.

CLIMATE MIGRATION AND TYPES OF CONFLICTS

Large-scale cross-border migration can induce conflicts in various ways. For instance, conflict can arise between receiver and sender states, or between the local population and the migrants, regardless of the underlying causes of migration. The expected scale of climate-induced migration is making the potential conflicts connected to migration a perceived threat to national security. Undoubtedly, the link between climate change, migration and conflict should not be simplified as there may be various reasons why conflicts would or wouldn't arise as a consequence of climate-induced migration. However, large-scale cross-border migration is increasingly perceived as a structural threat to many receiving countries. Particularly, if resources, jobs or economic opportunities are perceived to be threatened by the growing flow of migrants, growing tension can lead to conflicts which in turn may become violent, leading to forced deportation, loss of citizenship, and thereby risk denial of justice for migrants, especially those forcibly displaced (Swain, 2019). For instance, in October 2023, Pakistan enforced a law to deport all immigrants without legal status. More than one million Afghani immigrants were forced to go back to Afghanistan, even though several of them lived in Pakistan for decades. The majority of Afghanis who have left Pakistan fear that they will be arrested if they go back to Pakistan, according to the UN Refugee Agency and

International Organization for Migration (HRW, 2023). Although this case is not necessarily about climate migration, it sheds light on tensions that can arise between the receiving state and migrants.

Wherever displaced populations are forced to settle, they add to the local demands for food, water and other basic necessities. They also need to enter the job market. If there already is a lack of work opportunities, there is a prominent risk that migrants can put a greater challenge on host societies if they are not adequately prepared for a large inflow of people. Competition with local populations over shared resources can lead to conflicts between local and migrant populations. It can also produce political problems for the receiving state's regime as it may put pressure on the government (Jägerskog and Swain, 2024). Attempts by the receiving state to send the migrants back to their original country may incite tension between the two states (Swain, 1996a), similar to the recent dispute between Afghanistan and Pakistan. When experiencing resource scarcity in the new area, increasing feelings of nationalism among the local population of the receiving state can be generated. This can increase the organisation of the local population as a means to protect their own interests with the rationale that they, holding the nationality of the country, rightfully exist within their own, original, territory. The migrants, on the other hand, have other homes to which they can return (Swain, 2019). As a consequence, conflicts between the two groups can arise. In addition, the fear of migrant groups being radicalised also brings a perception of migrants as a security threat to the host society, especially when pointing out failures to integrate (Cochrane, 2015). These kinds of migration-related tensions and conflicts are not specific to climate migration. However, as climate change risks forcing more people to migrate, there is a prominent risk that the world may see an increase in these kinds of tensions and conflicts if political leaders are not prepared to deal with the emerging challenges.

Large-scale migration can also bring other kinds of conflictual situations. For instance, it can challenge the power dynamics among the elite of a country. As a response, the elite can actively build up a strong group identity within the receiving community, and incite groups to act against each other. By doing this, the elite can safeguard their interests. For instance, the elite may use and highlight ethnic differences between migrants and local populations to mobilise the groups to act against each other in fear of takeover by the other group (Swain, 2013). This type of conflict often stems from a feeling of insecurity among the elite, both within the migrant group and the local group. Consequently, it is an attempt to protect their own interests against each other (Swain, 2019). Another possible conflict scenario is that migrants may engage in demonstrations and anti-regime activities against their home country's governments, after being settled in a host country. Protection in the host society may enable an environment where migrants try to take revenge against their home country's regime, whom they may perceive as being responsible for their forced displacement.

For instance, through big demonstrations and protests. This is particularly common among conflict-induced migrants (Swain, 2019). However, as countries fail to ensure protection from the consequences of climate change for their populations, similar protests may occur for political climate inaction.

Commonly, international aid organisations host forcibly displaced migrants in camps in rural areas. In these camps, they are provided food, shelter, legal processing, education and medical care. However, migrants may prefer to resettle in urban areas where there may be a better chance of finding jobs and other opportunities, such as Cairo and Tunis which are expected to attract a large number of migrants experiencing water scarcity in their home locations (Clement et al., 2021). Declining rural economy through failures of agriculture and ecosystems can be another factor that induces migration to urban areas, within or across borders (Jägerskog and Swain, 2024). Rising urban migration is a challenge that national and international agencies are not well prepared for (Swain, 2019). For instance, rural migrants might end up living in human settlements with poor infrastructure where they are particularly at risk for natural hazards connected with climate change, such as landslides, pollution, access to water and flooding (UN-Habitat, 2023). Rapid urbanisation can create various social problems and can bring a large number of unsatisfied populations into close proximity. In turn, it can incite tensions and conflicts within a community, or between communities.

Inevitably, large-scale migration can pose real or perceived threats to countries' national security. Though the various types of conflicts that can be incited by migration are not connected with climate change, in particular, it does not mean that climate change will not play a part. Instead, climate change risks pushing migration to unprecedented levels. In turn, societies may find themselves in these kinds of conflictual situations if political actions are unable to address the emerging challenges connected with large-scale migration.

CLIMATE MIGRATION, POLITICAL CRISES AND RISE OF POPULISM

Given the challenges that can arise between and within countries if large-scale migration is inadequately handled, political anti-immigration mobilisation may be sparked. Anti-immigration mobilisation as a response to international migration has become a major reason for the populist surge that is increasing rapidly worldwide. In these kinds of movements, migrants are projected as negative influences on the 'culture' of the local population. Nationalist parties in the West tend to use the word 'culture' to describe differences with migrant groups, instead of explaining the differences on 'race'. However, the difference is mere semantics for all practical purposes (Swain, 2019). In other words, it is used as a way to justify the distancing of the migration community from the local community. This kind of exclusionary form of nationalism creates vast challenges for the protection and integration of migrants in receiving countries. Climate migrants are of course not an exception. Although the future of the scale of climate migration is uncertain, nothing is pointing towards that it would change the rising populism in receiving countries.

Populist parties and nationalistic agendas are rising rapidly throughout Europe. Many populist charismatic leaders have become popular and powerful in Europe by promising to protect their nationals against the 'invasion' of foreigners, refugees and even other European citizens. The political elite in European countries had set a consensus on major issues such as immigration for more than a century. However, the latest year's increase in large-scale migration has significantly changed the political reality across the European Union. The so-called Migration Crisis in 2015, with almost 1.3 million migrants mostly from Syria, but also from Iraq and Afghanistan, shook the previous agreements in the Union (PRC, 2016). The large-scale migration to Europe led the path to the rise of anti-immigration political mobilisation and nationalism based on the projected negative influence migrants may have on the 'culture' of the European countries (Swain, 2019).

The belief that the local population's 'culture' is superior to the 'culture' of the migrants is common for supporters of anti-immigrant political parties in Germany, Sweden, France and the Netherlands (Stokes, 2018). These far-right nationalist groups in Europe have been collaborating among themselves in an attempt to consolidate their power across the European Union. The populist and openly anti-immigrant politicians who once were seen as fringe extremists have moved into the political mainstream, though many of them are still in opposition. The growing nationalist and anti-immigration movements have led to policies that restrict the acceptance of migrants fleeing from war and violence, despite being obliged to provide security under international human rights law (Swain, 2019).

With increasing international migration from one country to another, more and more people are voicing their opposition to it. For instance, in 2018, a Pew Research Centre survey of 27 countries suggests that there has been a large drop in support for any type of migration (Connor and Korgstad, 2018). The declining public support for migration has encouraged nationalist parties and their leaders to use migration as an issue to strengthen their electoral position. In fact, framing immigration as a political issue that threatens the country's culture, religion, security, economy and politics seems to play a significant part in the electoral success of nationalist parties (Mudde, 2012). This kind of narrative is not only limited to Western countries. It has started to become a major political issue also in many migrant-receiving developing countries, such as India, Pakistan and South Africa (HRW, 2023; Myeni, 2022). As migration becomes a growing topic of debate, many countries are increasingly being pressured to make rules and regulations which discourage migration, or only allow one specific type of migrants that they want to receive. More restrictive border control is one of the most commonly used policy responses in most migrant-receiving countries (Swain, 2019). The growing opposition to migration is often considered racism by migrants, while the host society prefers to frame it as protecting their national interest (Jack, 2016). This kind of nationalism is a challenge for countries and international organisations to restrict and manage migration, while there is a massive increase in the number of people who willingly or forcefully migrate (Swain, 2019).

The perceived threats of large-scale migration have been used as an argument for anti-migration political parties to increase their mobilisation in different parts of the world. There is a prominent risk that unprecedented climate migration can be used as a political pawn by nationalist and populist forces, which could lead to even more conflicts and injustices. Unfortunately, the world is witnessing increasing nationalism in many migrant-receiving countries, and, at the same time, there is a global trend of decline of democracy. The dilemma is that migration as a threat is used as a narrative by nationalist political forces which can make them grow in size. In turn, growing nationalist political forces can undermine the democratic systems in many countries and the possibility for migrants to avail justice and equality in their new society, especially if they are being targeted as a threat to the country.

DEBATES AROUND THE USE OF THE TERM 'REFUGEE'

Given the above-mentioned challenges related to large-scale migration, climate migrants are a particularly vulnerable group of migrants as they often are forced to migrate, but are currently not protected under international law. Many attempts have already been made to conceptualise this phenomenon. Climate-induced migrants have been called 'environmental refugees', 'climate refugees' and 'ecological refugees' among other things. However, the accuracy of using the term 'refugee' for people who are displaced by non-political factors is debated. Currently, climate-induced forced migration is not included in the definition of a refugee as established under international law, which is the most widely used instrument providing the basis for granting asylum to persons in need of protection. The most widely used definition of a refugee is based on the 1951 Refugee Convention which describes a refugee as:

> someone who is unwilling to return to their country of origin owing to a well-founded fear of being persecuted for reasons of race, religion, nationality, membership of a particular social group, or political opinion. (UN General Assembly, 1951:3)

The convention was concluded at a time when climate migration was yet not a topic on the agenda. In more recent definitions of refugees, such as from the Oxford Dictionary, persons who are forced to flee to escape natural hazards are addressed (Behrman and Kent, 2022). However, this definition does not provide international legal protection.

Despite the lack of international legal protection for climate migrants in terms of refugees, courts have started to consider climate change or environmental factors in asylum processes. In this context, the 2015 ruling of the Supreme Court of New Zealand is quite significant. Ioane Teitiota from the Republic of Kiribati applied for asylum in New Zealand stating that his right to life was threatened due to sea-level rise which reduced habitable land in his country. He claims that the reduced available land spurs violent land disputes. Additionally, the man claimed that environmental degradation and saltwater intrusion contaminated freshwater leading to limited access to the clean drinkable water. Though the top court of New Zealand recognised the genuineness of a Kiribati man's contention of being displaced from his homeland due to sea-level rise, it could not grant him refugee status, reasoning that he wouldn't face prosecution if he returned home. After that, he was deported to Kiribati in 2015 with his wife and children. Ioane Teitiota then approached the UN Human Rights Committee (HRC). The HRC claimed that slow and sudden environmental changes can lead to population displacement where protection in a receiving country may be necessary.

Nevertheless, they concluded that it may take more than a decade for the effects of sea-level rise to be felt in the country. In other words, the Republic of Kiribati has time to implement appropriate measures to protect the population, potentially through safely relocating people (HRC, 2020). The decision left room for future claims of asylum as a consequence of climate change but still indicated that it can be difficult to prove that it seriously threatens the right to life. Though many called the decision of the HRC a landmark one, Ioane Teitiota continues to live in Kiribati (Behrman and Kent, 2020). This is the first decision of the UN Human Rights Committee on an asylum case on the grounds of climate change, but it has no jurisdiction to rule on refugee status (HRC, 2020). Nevertheless, the case brings a small piece of hope that the future could bring possibilities to get legal protection in another country due to climate-induced environmental changes that force people to migrate.

One of the arguments against using the term 'refugee' for people who are forced to leave their home location due to climate change is that climate-induced environmental changes may not be the only reason for someone to migrate, even though it can be a vital deciding factor. However, as discussed at the beginning of this chapter, migration is never a straightforward lifestyle decision. Behrman and Kent (2022) claim that even in the 1951 Refugee Convention's definition of a refugee, the well-founded fear of persecution is not described as the *only* reason for someone to leave their country of origin. Instead, it indicates that it is one of the factors for migration to take place.

Similarly, they deem that climate change might not be the only reason why someone migrates, but that it is one of the reasons for migrating (Behrman and Kent, 2022). In other words, they argue that forced climate-induced migrants should have the possibility to be considered refugees under certain conditions.

Currently, instead of referring to climate-induced migrants as refugees, the term 'climate migration' is used more frequently (Bettini et al., 2017). Climate migration is a broader term which allows for a nuanced understanding of climate-induced migration, where it is not necessarily connected with being forced to migrate due to threats to lives and livelihoods. Instead, migration can be understood as one of many adaptation strategies to climate-induced environmental change. It puts greater emphasis on people's ability to cope with emerging environmental challenges and to migrate when environmental conditions deteriorate (Bettini et al., 2017). In other words, strengthened resilience and adaptation in the home location can increase people's ability to choose adaptation strategies, which may be migration if other strategies fail (Methmann and Oels, 2015). In this context, migration as an adaptation strategy is undertaken to reduce vulnerability rather than being an automatic response to a natural hazard (Caretta et al., 2023). At the same time, if adaptation strategies fail to strengthen resilience, and the possibility of migrating is reduced because of the inability to pay for the cost, people can become trapped in environmentally degraded areas (Vigil et al., 2022; Zaveri et al., 2021). Eventually, the trapped populations risk facing even more severe human impacts and displacements by future natural hazards (Foresight, 2011).

Although migration in certain situations can be understood as an adaptation strategy to reduce vulnerabilities, the term 'climate migration' does not fully capture states' responsibilities to ensure that the basic rights of forcibly displaced migrants are fulfilled in their new location, especially if the migrants cross national borders. In particular, it does not capture the connection with justice that climate-induced migration arguably is connected with. Mainly, because while strengthening resilience and adaptation strategies to reduce vulnerabilities to climate change are crucial, they might not always be enough to prevent environmental degradation to the point where people no longer can sustain themselves and therefore are forced to migrate. Many of the people who are forced to or choose to migrate due to climate-induced environmental changes are some of the ones least responsible for the vast greenhouse gas emissions that contribute to anthropogenic climate change (Bettini et al., 2021). On the other hand, many countries that are significantly more responsible for climate change, like many countries in the Global North, will probably become less affected by climate-induced environmental changes that push people from their home locations. At the same time, even though climate-induced migrants would be recognised under international law as refugees, it does not necessarily mean that the highest emitting countries, currently or historically, are the ones that will protect them. Therefore, it does not necessarily signify that the highest emitting countries will give justice to people who cannot sustain themselves anymore due to climate change. However, including climate refugees protected under international law is a necessary step to ensure that forcibly climate-induced displaced people have the right to seek refuge in another country than their country of origin.

Currently, without legal protection from the international community, many of the forcibly displaced by natural hazards end up being internally displaced persons (IDPs). The number of IDPs has been growing consistently since the Norwegian Refugee Council's IDMC began documenting them over the last 20 years (IDMC, 2022). The growing number of IDPs is of serious concern because providing them aid and assistance can be difficult and a potential return process may not be as smooth and assured as anticipated. Even though the IDPs are larger in number and more vulnerable, there is a lack of focus and coordination among the international agencies to help them (Swain, 2019). The service access to IDPs in the camps is in general much worse than what is available for the locals outside the camp. IDPs are not only highly vulnerable economically, socially and psychologically, but their situation is politically precarious compared to even refugees. They continue to live under the jurisdiction of the same regime, which is directly or indirectly responsible for forcing on them the decision to migrate. In other cases, the states are weak and unstable or lack proper planning and resources to protect the IDPs. The home country continues to be the primary protector of the IDPs, and the international community's role is only complimentary. Additionally, IDPs and refugees living in camps are vulnerable to extreme weather events such as floods, which sometimes cause a third displacement (IDMC, 2021b).

Forced displacement can cause trauma and loss of dignity, affect mental health and damage social cohesion and available support structure. Climate-induced floods, droughts, heat and cold waves are also known to cause increased anxiety, depression and stress (Behrman and Kent, 2022). Though mental health consequences of climate change don't occur in isolation, it is, however, important to realise the extent to which climate change impacts human health and well-being, and this is currently not addressed in any international migration agreement, particularly not for IDPs.

IDPs are undoubtedly among the most disadvantaged people and are often not treated as well as refugees. They have less legal and physical protection and a less certain future. Sadly, not only does the international community continue to overlook the IDP crisis, but there is also minimal research and policy analysis on IDPs. There have been dozens of research institutes, policy think tanks and advocacy agencies in the Global North on refugee issues. Many more have come up in Europe in recent years, particularly after the 2015 refugee crisis. However, the interest in studying and analysing IDP issues is very limited and even rare. One of the primary explanations could be that almost all IDPs live in the Global South (Swain, 2019). The majority of them are in Colombia, Ethiopia, DRC, Sudan and Syria. Ukraine joined this list only when its war with Russia started in 2022. In other words, the Global North has seen the IDP crisis as someone else's problem in far-away places, unlike refugees who have become a significant factor in their countries' political polarisation. At the same time, climate-induced migration is also understudied in certain areas, as current research is quite geographically limited. Research on climate migration is overly represented in five countries, namely India, Bangladesh, Ghana, China and Ethiopia, despite evidence of extreme weather events such as drought, floods and extreme temperature triggering migration in large parts of Africa, South Asia, Southeast Asia and Central America (Caretta et al., 2023). Undoubtedly, there is a need to advance research and policies on climate migration in many parts of the world, including for IDPs.

OVERVIEW OF EXISTING RESPONSES

Although climate migration has until recently been viewed as a peripheral concern, there is an ongoing collective global effort to address international migration in which climate migration is being included. Giving hopeful signs, the Global Compacts have been affirmed by the UN General Assembly, and a United Nations Framework Convention on Climate Change (UNFCCC) Task Force on Displacement has been created. The Task Force of Displacement has the goal to develop recommendations to avert, minimise and address climate displacement. The Global Compacts are non-binding political declarations that address climate migration and are significant milestones in getting a broad global consensus on collective action on forced migration-related challenges. Nevertheless, no legal rights for climate refugees are created through them (Behrman and Kent, 2022). The Global Compact on Refugees provides an outline for international organisations and other key stakeholders to support host countries and communities, help refugees lead productive lives and work for the safe and dignified return of the refugees. The Global Compact for Safe, Orderly and Regular Migration was affirmed on 19 December 2018, focusing on the shared responsibility of the international community to address international migration jointly. While the Global Compacts have opened up the discussion about climate migration, neither full recognition is given to them nor strengthened protection for increased support to current and future climate migrants (Behrman and Kent, 2022). Nevertheless, the Global Compacts are a major landmark in getting a broad global consensus on migration-related challenges. Cooperation between and among the migrant-sending and migrant-receiving countries is critical for the migration process to be smooth and less conflictual. Like the Global Compacts at the international level, regional organisations need to develop a common policy framework on forced migration in particular and migration in general, which should be clear, transparent, humane and sustainable.

Being forced to migrate in search of survival is never easy. There is no doubt that climate change will continue to force millions of poor people to migrate, and many of them will cross national borders for survival. Though there is an ongoing debate regarding whether climate migrants should be defined as climate refugees, climate migration is yet to be included in the definition of a refugee as established under international law. In other words, 'climate refugees' continue to lack the legal protection that provides a basis for granting asylum. To date, no refugee convention or multilateral treaty has provided legal protection for climate-forced migrants. They are only covered through human rights law. There is a need for the definitional fiat of 'refugee' to be expanded to address the increasing challenge of climate change-forced displacement. The increasing threat of climate-induced forced migration needs to be taken seriously by the international community and regional organisations. A coordinated effort at the global and regional level only can motivate, coordinate and implement an effective approach to address the unavoidable climate-forced large-scale population migration.

When the number of environmentally or climate-induced displaced people is more than 10 million every year on average, and also, most often, the level of suffering is as terrible as conflict-forced migrants, they simply cannot be ignored by the multilateral institutions. In relation to the increasing migration-restriction policies in the world, it is important to address and act on the responsibility of countries contributing most to climate change (Bettini et al., 2017). While it is crucial to implement adaptation and mitigation measures to improve resilience towards climate change, it is equally important to ensure the protection of the rights of the people who are climate-induced forcibly displaced. International refugee agencies in the past have not been able to save the lives of many environmentally displaced people in the Global South due to the absence of their mandate. This fast-emerging crisis demands an urgent relook at the conceptual use of the term 'refugee' and engaging in a sincere and coordinated effort to make the necessary adjustments to include climate-forced migration. Ultimately, finding an equitable balance between the number of people and available natural resources should be the rationale reasoning behind achieving long-term commitment, a better early-warning system, incorporation of the migration variable in state planning and acceptance of positive influences of migration (Swain, 2019).

SUMMING-UP

The phenomenon of the escalating security challenges posed by climate-induced migration has transformed from a peripheral concern to a major international issue. The dramatic effects of climate change, ranging from sea-level rise to extreme weather events, have led to unprecedented levels of migration. These movements are not only due to direct environmental changes but are also influenced by complex socio-economic and political factors. The scale of climate-induced migration is vast and varied, impacting both sudden and gradual environmental changes. It has become evident that traditional frameworks for understanding and managing migration are inadequate in the face of this growing challenge. The lack of legal recognition and protection for climate migrants under international law, as exemplified by cases like Ioane Teitiota's, highlights the urgency of the situation. The use of terms such as 'climate refugees' remains contentious and legally non-binding, leaving a significant gap in the protection of these vulnerable groups.

The international community faces numerous challenges in accommodating climate migrants. The rise of nationalism and populism, fuelled by anti-immigration sentiments, complicates the response to this crisis. As climate-induced migration increases, it has the potential to exacerbate existing tensions and create new conflicts, both within and between nations. The strain on urban infrastructures due to migration to cities, the competition for resources and the alteration of power dynamics within countries further add to these challenges. The plight of IDPs, who often face even greater vulnerabilities, cannot be overlooked. The international community's current response, while a step in the right direction, falls short of the urgent need for a comprehensive and coordinated approach to address the complexities of climate-induced migration.

There is an urgent need for a significant rethinking of international policies and legal frameworks to better protect and accommodate climate migrants. A global effort is essential to redefine the concept of refugees to include those displaced by environmental changes. Moreover, there is a pressing need for more robust, collaborative strategies that go beyond mere acknowledgement of the problem, focusing instead on concrete actions to mitigate the causes of displacement and ensure the rights and dignities of climate migrants. Only through a concerted and empathetic global response can the challenges posed by climate migration be effectively addressed, ensuring security and justice for all affected populations. Thus, the climate migration should be a national security issue for the governments not in order to create an island for themselves, but to see it as an issue that has to be dealt in cooperation with other countries adhering to international law and humanitarian norms. That is the only way they can address the issue of this magnitude.

7
REGIME LEGITIMACY AND DEMOCRACY

This chapter embarks on a critical exploration of the multifaceted challenges posed by climate change, a crisis that transcends geographical and political boundaries. At the heart of this global conundrum lies a glaring disparity: the Global North, despite its significant contributions to greenhouse gas emissions, has shown a troubling lack of concrete action in mitigating climate change. Conversely, the Global South, grappling with its own unique socio-economic constraints, often finds itself mired in inadequate planning and engagement, struggling to effectively confront the environmental upheavals. This dichotomy not only exacerbates the global climate crisis but also casts a shadow over the political landscape. Non-democratic regimes face mounting pressures as climate change tests their governance and resilience, while even the most stable democratic institutions are not immune to the destabilising effects of environmental calamities. The potential of climate change to undermine regime legitimacy and destabilise established political orders presents an urgent call to action, necessitating a cohesive and far-reaching response from the international community.

THE IMBALANCE BETWEEN RESPONSIBILITY AND CONSEQUENCES

The climate crisis affects all parts of the world, but some countries suffer more than others. While the Global North is primarily and historically responsible for the climate crisis, the Global South suffers the most from its consequences. According to the Global Climate Risk Index, the countries most affected by climate risks are Mozambique, Zimbabwe, the Bahamas, Japan, Malawi, Afghanistan, India, South Sudan, Niger and Bolivia (Eckstein et al., 2021). The index ranks countries based on the frequency and intensity of weather-related loss events experienced during the year. This index has not been updated between 2019 and 2023 because of a temporary lack of data (Germanwatch, 2023). However, the data from 2019 reveals that during a longer perspective, between 2000 and 2019, the most affected countries/territories were Puerto Rico, Myanmar, Haiti, the Philippines, Mozambique, the Bahamas, Bangladesh, Pakistan, Thailand and Nepal (Eckstein et al., 2021). Besides Japan and possibly Puerto Rico, all other top ten countries are struggling to develop economically and politically.

The University of Notre Dame in the United States has compiled another index ranking countries most affected by climate change, through the Notre Dame Global Adaptation Initiative (ND-GAIN). This index also illustrates that it is mostly countries struggling economically and politically that are most at risk of facing severe climate-induced challenges. The ND-GAIN index does not only take vulnerability to climate change into account but also the possibility for a country to adapt to the changing climate. In other words, the country's capability to strengthen its resilience. Mainly, because depending on the economic strength and governance structures, countries can respond to and be affected by climate challenges differently (Busby, 2022). Therefore, some countries are better at planning and implementing mitigation and adaptation strategies.

According to the ND-GAIN Country Index, the top five countries with low vulnerability and high readiness to deal with climate challenges in 2021 were Norway, Finland, Switzerland, Denmark and Singapore (ND-GAIN, 2023). While these countries are vulnerable to the changing climate through faster warming temperatures than the global average in Scandinavia, melting glaciers and moving borders in Switzerland and rising sea-levels in Singapore, these countries' readiness to deal with these challenges is comparatively high, although expensive (Harris, 2016). All these countries are high-income economies. In fact, out of all the 185 ranked countries in the ND-GAIN Country Index, the top 60 countries are high-income or upper middle-income countries. The majority of these countries are located in Europe. Nevertheless, Europe is one of the fastest-warming continents on the planet. The continent has been warming twice the global average since the 1980s (WMO and CCCS, 2023). The Arctic, which partly is located in Europe is warming almost four times faster than the global average (Rantanen et al., 2022). Although traditional livelihoods and ecosystems in the region are highly threatened by the warming planet, Europe, together with North America, are the most well-equipped countries to adapt to climate change by strengthening their resilience. These temperate northern regions face the highest relative change in temperature increase, compared to countries in tropical and subtropical regions. However, countries located in the tropical and subtropical regions that already experience warmer temperatures are largely at risk of extreme heat events even with smaller temperature changes. In particular, countries with high levels of humidity are at risk of becoming most severely affected by extreme heat as humid conditions aggravate human's heat-coping abilities (Zhang et al., 2023a). Therefore, it is not the places where the temperature rise will be the highest relative to pre-industrial times that will suffer the most from its consequences. Particularly, because many countries in the tropical and subtropical regions in the Global South generally have fewer resources to adapt and mitigate the consequences of climate change, partly because of their lower economic development (Harris, 2016).

According to the ND-GAIN Country Index, the countries that were most vulnerable and least prepared to deal with the changing climate in 2021 are Chad, Central African Republic, Eritrea, Democratic Republic of Congo and Guinea-Bissau (ND-GAIN, 2023). All five countries are located in the African continent and suffer from weak economies and political instabilities. In addition, they have contributed very little to climate change while also having minimal say in climate negotiations (IPCC, 2022b). Not only because they lack economic and political heft, but they also don't have enough qualified negotiators to engage in the compound climate negotiations. These nations, often at the frontline of climate change's harshest impacts, are hindered by a dearth of experts who can adeptly navigate the labyrinthine complexities of climate diplomacy. This deficit is not merely a matter of numbers; it reflects a profound disparity in access to specialised training, resources, and the cumulative experience often abundant in more affluent states. As a result, these countries frequently find themselves on the periphery of crucial discussions, their voices and concerns diluted in the chorus of global debate and compromises. Their struggle to field seasoned negotiators means that the intricacies and nuances of climate agreements might bypass their interests, leaving them at a disadvantage in a scenario where every decision shapes the future of their vulnerable communities. In general, most countries in Africa are disproportionally affected by the changing climate. Out of the 50 most vulnerable and least prepared countries, 33 countries are located on the African continent. The other 17 countries are located in South America, the Pacific Ocean and Asia (ND-GAIN, 2023). It is a cruel irony that people and countries in these regions are already suffering the most from climate change but have done the least to cause it. In fact, the 52 poorest countries of the world have contributed less than one per cent of global carbon emissions (Our World in Data, 2021). Conversely, the G20 countries, a collection of the world's largest economies, are responsible for almost 80 per cent of the global greenhouse gases.

Undoubtedly, all countries are suffering from climate change, but countries in low-income countries are disproportionally suffering from its consequences. Widespread poverty, poor governance, weak institutions, corruption and lack of technologies are aggravating the vulnerability of countries with developing economies. In addition, many of the most affected countries are conflict-affected and therefore fragile to the additional burden added by climate change. Conflict-affected countries often experience social fragmentation through social and ethnic tensions which obstruct efforts for climate and sustainability actions. In other words, conflict-affected fragile societies undermine the state's ability to adapt to the new situation emerging out of climate change.

CLIMATE INSECURITY MEANS ECONOMIC INSTABILITY FOR THE GLOBAL SOUTH

As the climate crisis accelerates, economic growth and development gains risk reversing, particularly in the Global South. Mainly, because poorer countries in the Global South lack the economic strength to ensure the implementation of effective adaptation measures. In turn, climate change can make livelihoods less secure in various ways by posing several unconventional security challenges. Climate change brings unprecedented societal and economic security challenges relating to food, water and energy security. The health and well-being of people and ecosystems are also threatened. In addition, the climate crisis brings an amplified risk of an increasing number and intensity of new and already existing conflicts, resulting in more devastating humanitarian crises. As many countries in the Global South suffer from poor governance and weak institutional structure, they have limited capability to deal with these unprecedented challenges. Populations all over the world face more extreme weather events as a consequence of global warming. However, the consequences in the Global South are often more severe than in the Global North. In the Global North, states often have the resources and capabilities to ensure sufficient recovery after extreme events. In many countries in the Global South, resources often exist for acute rescue and relief operations, but the economic strengths to fully recover and rebuild from the events are more precarious.

One of the most effective ways of adapting to climate change and protecting people and properties from natural hazards such as storms, floods and heatwaves is early warning systems. Early warning systems help communities prepare for climate-induced natural disasters and allow time for evacuation. These kinds of systems have proven to substantially reduce climate-induced fatalities and economic losses. However, only half of the countries worldwide have implemented these early warning systems, according to the World Meteorological Organization (WMO, 2021b). In particular, Small Island Developing States (SIDS) and the least developed economies lack systems that protect people and livelihoods by forecasting severe weather such as flash floods, coastal inundations and hurricanes. In turn, the population in these countries are more at risk of being severely affected by climate-induced natural hazards.

Globally, between 1970 and 2019, there have been 11,778 natural hazards leading to two million deaths and US$4.3 trillion in economic losses. Over 90 per cent of all the reported deaths have been in countries with developing economies located in the Global South. On the other hand, countries in the Global North have experienced 60 per cent of the reported economic losses from natural hazards. However, the economic losses in the Global North have been less than 0.1 per cent of the countries' gross domestic products (GDP), respectively. In contrast, in the least developed economies, seven per cent of all disasters lead to economic losses greater than five per cent of their GDP. The SIDS have been most affected by economic losses due to natural hazards in terms of the percentual loss of GDP. In the SIDS, around 20 per cent of the natural hazards have led to an impact exceeding five per cent of the GDP (WMO, 2021b). In some cases, severe natural hazards have led to economic losses above 100 per cent of the GDP. For instance, the devastating Hurricane Maria swept over the Lesser Antilles, Puerto Rico and the Bahamas in 2017. In the small island state of Dominica, the hurricane caused damages of US$ 1.31 billion, equivalent to around 200 per cent of the country's GDP. The hurricane destroyed almost all trees and vegetation and eliminated the agricultural sector. When Hurricane Maria hit, the country was still recovering from the effects of Storm Erika which during 2015 caused GDP losses of around 100 per cent of the country's GDP (IMF, 2019). Undoubtedly, small countries experiencing reoccurring natural hazards are particularly at risk. Especially, if they don't have early warning systems which could help them prepare for the upcoming extreme weather. As the economies of poorer countries in the Global South are generally more negatively affected by climate-induced natural hazards, climate change risks widening the North-South divide further.

Inevitably, climate-induced weather extremes have caused devastating societal effects in countries in the Global North as well. For instance, Germany experienced severe flooding in July 2021. Almost 200 people lost their lives and the infrastructural damages went up to a cost of US$34 billion (Koks et al., 2022). It was the country's highest number of fatalities and infrastructural damages by natural hazards on record (Cornwall, 2021). Similarly, in 2023, Norway experienced infrastructural damages of the highest cost ever, when the country was hit by the severe Storm Hans. The storm caused damages of almost US$ 170 million (Rising, 2023). During the storm, thousands of people were evacuated from their homes as they were warned about heavy precipitation and rising water levels in the rivers. Moreover, the United States frequently experiences deteriorating hurricanes that cause massive destruction and many fatalities. In 2018, Hurricane Michael hit Florida. It was the strongest hurricane on record and caused US$18.4 billion in damages. While some low-income households have not fully recovered yet, recovery funds from insurance companies and grants and loans from governmental agencies were made available for the population to recover from the effects at large (Moens, 2022). All three countries have strong governance systems and institutions that facilitate the recovery from extreme weather events. Nevertheless, the recovery period can take time and is costly. In addition, sometimes disadvantaged sections of society can be more vulnerable and not get sufficient help to fully recover. However, the general societal effects of extreme weather events in the Global South can lead to even more devastating short-term and long-term consequences.

Extreme weather events such as droughts, flooding, storms and hurricanes risk aggravating already existing societal problems. For instance, in 2022 and 2023, the Horn of Africa suffered from its worst drought in 60 years (WMO, 2023c). The drought has put 23.4 million people in acute food insecurity, leaving 5.1 million children acutely malnourished and 2.7 million people displaced (WFP, 2023). The extreme drought has severely destabilising impacts on the societies and obstructs the ability to ensure development gains for their populations. The drought has deteriorating effects on agricultural and pastoralist livelihoods. It has also increased health-related vulnerabilities, reduced access to education and gender inequalities (Acaps, 2023). In other words, climate change can seriously exacerbate the social impacts of natural events by increasing the intensity and frequency of weather extremes.

Currently, the education of 40 million schoolchildren is disrupted annually through natural hazards that destroy infrastructure, degrade learning environments and increase poverty rates (Wood, 2023). The world has committed to ensuring quality education for everyone by 2030, as a part of the 17 Sustainable Development Goals (SDGs). However, following the current path, 84 million children worldwide will not attend school, and 300 million students will lack basic numeracy and literacy skills by 2030 (UN, 2023d). The inability to adapt and mitigate the consequences of climate change exacerbates the ability to reach this goal. For instance, in Zimbabwe, heavy precipitation in 2016 and 2017 destroyed 18 per cent of the country's schools, disrupting the education of 500,000 schoolchildren (UNDP, 2017). The disruption of basic societal systems, such as education, can have devastating long-term effects on poor households and countries which can obstruct the possibility to strengthen their economic situation to ensure their well-being.

The educational disruptions also risk deepening gender inequalities as families in poverty often prioritise sending their sons to school if they cannot afford to educate all their children. In low and lower middle-income countries, agriculture is the most important employment sector for women. They are also often responsible for fetching water and preparing the food. During periods of droughts or floods, women usually have to work harder to ensure income, water and food for their families. Girls can be pressured to leave school to help their mothers in the field (UN Women, 2022). In addition, climate change effects on poverty and education risk exposing more girls to child marriages as a way for families in poverty to deal with the situation (UNDP, 2017). Women and children are also at risk of increasing climate-induced health challenges. For instance, extreme heat risks increase the cases of stillbirth. Global warming is also expected to accelerate the spread of Malaria, Dengue fever and Zika virus which risk worsening childbirth complications and children's health (UN Women, 2022). Moreover, increased frequency and intensity of climate-induced natural disasters risk reducing physical and mental health, increasing population displacements and exacerbating already existing inequalities (Wood, 2023). There is no doubt that climate change endangers human health and well-being in multiple ways affecting all sectors of society and at all levels, particularly in poorer economies in the Global South. In turn, this can harm social cohesion and community resources (APA, 2023). Inevitably, the consequences of climate change in low and lower middle-income countries bring larger implications than only infrastructural damages and fatalities. Instead, it risks deepening the existing inequalities and worsening the societal problems.

The livelihoods and incomes of households, communities and countries can negatively be affected by climate-induced natural hazards, weather extremes and changes in climatic patterns. The ones most at risk are the poorest people and countries. The world has in recent years experienced an increase in people being pushed back into poverty. At the end of 2022, almost 685 million people lived in extreme poverty (World Bank, 2022c). Climate change is expected to push even more people into poverty. The World Bank estimates that by 2030, an additional 68 to 135 million people risk being pushed into poverty as the intensity and frequency of extreme weather events accelerate. Sub-Saharan Africa and South Asia are particularly at risk for climate-induced poverty, as these regions already account for the largest share of poverty rates (World Bank, 2020). In addition, climate change is also expected to negatively impact employment rates. The International Labour Organization (ILO) estimates that heat waves will result in the loss of 80 million full-time jobs worldwide, for a cost of US$2.4 trillion. South Asia and West Africa are expected to be hit hardest, losing the greatest number of working hours. Mainly, it will have the most effect on the agricultural sector, which means agricultural-based economies will be hit hardest (ILO, 2019). The effect on the agricultural sector is also likely to result in food shortages. In turn, the challenges of feeding growing populations while facing climate change and shortages of freshwater supply and arable land are becoming serious concerns worldwide, particularly in poorer countries.

In South Asia, where agriculture is central to the economy, climate change has made the occurrence of heatwaves 30 times more likely (Zachariah et al., 2023). Heatwaves can be extremely dangerous and are among the deadliest climate-induced natural hazards. Extreme heat's cascading effects on food production, human and animal health, energy generation and infrastructural capacity can seriously threaten the well-being of humans and ecosystems, while also leading to huge economic losses. It predominantly affects people in poverty as they often have no other choice than to provide for themselves by working outside in sectors that will be highly affected by climate change.

South Asian countries are highly affected by various natural hazards. For instance, in 2022, Pakistan witnessed unprecedented floods where at least one-third of the country was under water. Pakistan is one of the most high-risk climate countries. The country has more than 7,200 glaciers, more than any country in the world. The glaciers are melting faster and earlier in the season than normal and contributed to the devastating floods. On top of that, the country received a monstrous amount of monsoon rain in July and August 2022. The rainfall was the highest in the last three decades. On average the country received 190 per cent of the normal rainfall during these months. Some parts of the country even received 450 per cent more rainfall than normal (Nabi, 2023). The floods affected more than 33 million people. It destroyed two million homes, 13,000 km of highways and four million acres of agricultural land. As a consequence, millions of people were pushed into poverty, and eight million were displaced (Gallo et al., 2023; Nabi, 2023). The catastrophic flooding caused at least US$ 40 billion in damages, which will have devastating effects on the country's economy (Bushard, 2022). To avoid long-term effects on the country's economic growth and livelihoods of low-income households, the country received state assistance from international actors such as the United Nations, the World Bank, the Asian Development Bank and the European Union (Nabi, 2023). Besides Pakistan, the flooding in 2022 also affected India, Bangladesh and Nepal. South Asia is highly vulnerable to floods and is the region most vulnerable to physical risks associated with climate change (Harris, 2016). South Asian countries also suffer from poor governance, weak institutions and poor infrastructure, thus being less equipped to cope with and manage extreme flooding. Moreover, the major rivers in South Asia, like the Indus and Ganges and the Brahmaputra, which pose significant flood risks for the region, are transboundary. Unfortunately, there is very little cross-border cooperation among the South Asian countries over managing the floods of these rivers. In 2023, flooding in the northern part of India led to a political blame game over rescue and relief measures between the centre and states and between the government and the opposition. The costs of flooding are often massive, and it is predominantly the poorest part of the population that is mostly affected. What is new for the region is that extreme floods have become more frequent and devastating, causing much more damage than ever before.

Undoubtedly, many regions and societies have been left behind and struggling for survival as they are becoming increasingly affected by climate-induced natural hazards and extreme weather. At the same time, the quest for continued economic growth, and in some cases, even providing food and shelter to a rapidly expanding world population has led to widespread devastation of renewable natural resources like land, air, water, forest and biodiversity, particularly in the Global South. The decline of these resources already threatens the life and survival of present and future generations. Climate change has brought further uncertainties over the availability of these critical resources and even threatens the survival of this planet.

While all countries are affected by climate change, it is clear that the societal consequences in countries in the Global South generally are more severe than the consequences in the Global North. Development gains are larger at risk in countries in the Global South.

The economic burden of climate change disproportionately affects the most vulnerable populations, perpetuating existing economic inequalities. In addition, the demand for international aid and assistance to cope with the challenges of climate change is more prominent in the Global South as their economies are weaker, have put less effort into adaptation measures and therefore may be more severely affected. In general, countries with weak state capacities and political institutions are most at risk of exacerbated climate-driven insecurities and violence. Mainly, because these countries lack the needed bureaucratic and technological strength to solve national problems and provide support in humanitarian emergencies. Professor Josh Bushby, who was a senior advisor for climate in the US Department of Defense, writes that climate-driven insecurity challenges risk increasing if countries fail to provide expected and demanded rescue, relief and rehabilitation assistance (Busby, 2022).

CLIMATE RISK REGIONS: FERTILE GROUND FOR AUTHORITARIANISM?

Clearly, the universal nature of the climate crisis underscores that it is not only an environmental issue but it has also become a major driver of economic inequalities, political instabilities and societal divisions globally. Climate change worsens existing vulnerabilities, widens the gap between privileged and marginalised groups and can trigger disputes over resources and migration. Ultimately, climate change as a 'threat multiplier' can deteriorate democracies worldwide. Currently, many democracies in the Global South struggle to effectively address climate change. This can lead to economic decline, unemployment, food insecurities, growing injustices, climate-induced conflicts, migration and financial instability, which undermine the prosperity they promise their citizens. In turn, climate change can negatively influence democratic systems (Lindvall, 2021). Mainly, because the climate crisis places additional demands on states, which if unmet can undermine regime legitimacy.

As climate change exacerbates poverty and economic instability, democracies risk being negatively affected. Economic welfare and well-functioning welfare systems are often considered prerequisites to facilitate the development and maturing of a democracy (Lindvall, 2021). Income disparity can diminish mutual trust, and increase polarisation, criminality, corruption and social unrest (Mounk, 2018). Existing social disparities can also be exacerbated as climate change disproportionately affects vulnerable populations. In addition to economic and social consequences, climate change has serious implications for public health.

Democracies have a duty to deal with these societal challenges to ensure justice and protect the health and well-being of their citizens. Failing to effectively address climate change not only jeopardises public health and well-being but also results in increased healthcare costs and a lower quality of life, eroding the legitimacy of democratic systems further. These challenges can overwhelm societal systems like health care and social services, particularly in vulnerable communities (Lindvall, 2021). Moreover, climate-induced migration can create tension between displaced communities and host communities. In turn, it can erode trust and social cohesion. Competition for limited resources like water and fertile land often results in conflicts, putting democratic governments in difficult positions as they struggle to maintain control and provide security, ultimately undermining their legitimacy.

Social and political instability, along with diminishing trust in state institutions, pose significant challenges to maintaining order in many democracies in crisis. Many democracies, especially in the Global South, have been slow to respond to the climate crisis, and their efforts have often fallen short. This lack of decisive climate action erodes public confidence in their ability to address climate change challenges. Inaction on climate change not only threatens the environment but also undermines democratic ideals of justice, equity and representation. Especially if citizens perceive that their government prioritises corporate interests over the needs of the people, trust in democratic governance further diminishes. In particular, if countries fail to provide sufficient aid, relief and recovery for affected populations. In these societies, climate-induced hazards and weather extremes can trigger growing societal unrest and disturbances (Lindvall, 2021). More collapsed states may also require more international interventions for peacekeeping purposes (Swain and Öjendal, 2018). Thereby, being dependent on external actors, which in turn can spark further distrust in government legitimacy to deal with the emerging challenges. Ultimately, democracies risk reverting to populist autocratic governments.

Populism can be considered a threat to democracy as it can weaken and diminish democratic institutions. Populists risk increasing polarisation by blaming minority groups, traditional elites or other countries for sabotaging the country's security and its success. The arguments can be used to stay and gain further power, ultimately risking the democracies becoming authoritarian regimes (Müller, 2016). Authoritarian regimes are opposite to democracies and lack the core entities of a democratic structure. Under autocratic rule, state control is strong, political pluralism is lacking and political repression is common (Fiorini, 2018). Conversely, in liberal democracies, protection of the population, equality and justice are core principles.

While the world is facing the severe challenge of the ongoing impacts of global warming, it simultaneously faces the challenges of increasing populist autocratic governments in various regions. Although these two challenges are not necessarily connected, the negative effects of environmental disruption have profound consequences for the stability and prosperity of democracies worldwide. While the last century has provided more countries and people to enjoy democratic rights than previously in history, this trend has recently started to reverse. In 2016, 96 countries were classified as full democracies. In 2022, the number has fallen to 90 (Herre, 2022). In fact, it was the first year in two decades that the world had more closed autocracies than liberal democracies.

The database Varieties of Democracies (V-Dem) outlines in their 2023 report that only around 13 per cent of the world population lives in liberal democracies, whereas 28 per cent lives in closed autocracies in 2022. Mainly, there is a drastic shift in the number of countries going through autocratisation, moving away from democracy. In 2002, 43 countries were undergoing democratisation, moving away from autocracy. In the same year, 14 countries were autocratising. Conversely, in 2022, 13 countries are democratising and 42 countries are autocratising. Only in two years, between 2020 and 2022, nine countries have developed into closed autocracies. These are Afghanistan, Chad, Guinea, Haiti, Iran, Myanmar, Turkmenistan and Uzbekistan. Populations in countries going through democratic backsliding (i.e., decreasing democracy) experience weakened freedom of expression and association, stronger government censorship, a growing spread of disinformation and increased polarisation. The Asia-Pacific region faces the most dramatic democratic backsliding, with levels close to the ones recorded in 1978. In this region, almost 90 per cent of the population lives in autocracies. Mainly, the democratic backsliding in India with its 1.4 billion people accounts for a large part of the population living in autocracies. In fact, India has in the last ten years been one of the worst autocratising countries worldwide (V-Dem Institute, 2023).

Climate change is not necessarily connected to the democratic backsliding going on in several countries worldwide. However, as political legitimacy and trust are core concepts in democracies, democracies that fail to act on the emerging climate challenges, risk putting their own stability, prosperity and fundamental principles of justice and representation in jeopardy. As climate change accelerates and its consequences become more visible and devastating, governments worldwide face significant challenges to their legitimacy. In fact, in some cases, the frustration of inadequate responses to climate change-related challenges and perceived government inaction have made populations increasingly turn to populist authoritarian leaders, effectively shifting their countries towards autocracies. In these cases, authoritarian regimes can be perceived to better deal with natural hazards and extreme weather because they may respond more decisively than democratic leaders in weak democracies (Lin, 2015). In some countries in the Sahel region of Africa, military coups have overthrown struggling democracies, with citizens celebrating coup leaders as liberators. In fact, while the populations in strong democracies with high-quality institutions generally suffer the least from natural hazards and extreme weather, populations in weak democracies risk suffering more from the consequences of governmental inaction than populations in authoritarian regimes (Ahlbom Persson and Povitkina, 2017). Especially, if the ruling parties are mainly focusing on staying in power in the next elections, rather than focusing on ensuring the well-being of its population by meeting their demand (Fiorino, 2018). Therefore, democracies worldwide must recognise that climate change represents a survival crisis for them if they are unable to provide what is demanded by the populations (Busby, 2022).

For instance, post-disaster national survey data from the 2010 earthquake and tsunami in Chile showed evidence that the inability of the state to provide sufficient relief aid decreased the population's trust in governmental legitimacy. The natural hazards caused 500 deaths and around one million displacements. While the governmental trust weakened, the support for military coups increased (Carlin et al., 2014). Similarly, in the aftermath of the earthquake in Peru in 2007, support for political actors and democracy was undermined because of the political inability to respond to the citizens' demands (Katz and Levin, 2015). In countries undergoing economic transition, disappointing policy outcomes can erode citizens' trust in democracy and potentially to instability and even collapse. If the accelerating challenges connected to natural hazards are handled insufficiently, the governmental inaction risks facilitating a shift to authoritarian and populist governments, particularly in struggling democracies. Nevertheless, the concentration of power in authoritarian regimes can be problematic. Especially because autocracies restrict freedoms and rights and risk manipulating information regarding relief aid, affected people and climate actions (Lin, 2015).

In several small island states, reoccurring severe storms have been connected to increased and persistent autocratic tendencies ever since the 1950s (Rahman et al., 2022). Many storm-prone small island states have persistent authoritarian regimes, such as Haiti and Fiji. These countries have even been called 'storm autocracies' by researchers because of the connection between reoccurring storms and persistent autocratic tendencies. After storms and other natural hazards, affected populations require post-disaster aid assistance. In these kinds of states, the vulnerability of the population is taken advantage of by providing the necessary aid to avoid social unrest, but at the same time employing a stronger non-democratic orientation of the government (Rahman et al., 2022). In other words, climate-induced disasters can strengthen and legitimise the authoritarian grip on society if they ensure sufficient rehabilitation relief (Lindvall, 2021). The immediate demand is thereby fulfilled, but in return, the population becomes more repressed in the long term. Small island nations are particularly at risk for this because large parts of the country will be hit by the reoccurring storms. Therefore, it is easier to take control of the whole country, compared to if only a smaller part of the country is affected. Especially, if natural disasters are reoccurring phenomena (Rahman et al., 2022).

Moreover, food insecurity and rising food prices are also a threat to democratic systems as they can trigger societal unrest, urban riots, demonstrations and political instability. In fact, food insecurity and hunger substantially constrain the possibility for elected leaders to stay in power (Lindvall, 2021). Mainly, democracies with strong institutions and political stability are often better at dealing with food insecurities and connected challenges, compared to weak democracies. Strong democratic institutions therefore do not face substantial risk of breakdown when food prices surge. Instead, the leading parties risk not being re-elected if they fail to handle to situation as demanded by the citizens, but it seldom leads to threats to the democratic system itself (Lin, 2015; Rahman et al., 2017). Weaker democracies, on the other hand, risk going through democratic backsliding if the population don't have sufficient food to meet their needs. Between 1970 and 2007, food insecurities and rising food prices have been connected to many cases where democratic institutions have substantially deteriorated, and riots, civil disturbances and the rise of anti-government demonstrations have had an upswing in low-income countries (Arezki and Bruckner, 2011). Countries with weak democratic institutions are particularly at risk when experiencing food insecurity and increased food prices. In authoritarian regimes, social unrest as a consequence is less probable because of the power of the regime to 'silence' its population (Hendrix and Haggard, 2015).

The convergence of climate change and the rise of populist autocracies pose a serious threat to democracies worldwide. In turn, it also risks further constraining the efforts to limit global warming to protect our planet. To address these existential threats, democracies must take bold and decisive action to combat climate change and protect the well-being of their citizens by enhancing their climate adaptation plans. If governments take appropriate action, climate-induced challenges can facilitate democratisation processes, instead of autocratisation. For instance, the 2015 earthquake in Nepal is considered by some as having somewhat facilitated the democratisation of the country. The devastating earthquake caused 9,000 fatalities and severely damaged infrastructure throughout the country. At the time of the earthquake, the country was in the process of creating a new constitution. The aftermath of the earthquake contributed to bridging political opposition to respond to the new challenges. Two years later, in 2017, Nepal held its first election after 20 years. Although Nepal is a young democracy, the earthquake seems to have contributed to increased democratic tendencies (Pokharel et al., 2018). Clearly, it is possible to collaboratively act on the emerging challenges to strengthen democracies rather than undermining them. Therefore, democratic leaders must understand the challenges ahead and act to ensure that fewer countries will fall into autocratic regimes which is a threat to human rights at large.

Democracies, opposite to autocracies, are built on the idea of intergenerational responsibility, where each generation is tasked with ensuring the well-being of the next. Climate change serves as a stark example of how neglecting long-term challenges betrays this responsibility. Democracies that fail to take strong action on climate change prioritise short-term interests over future generations, contradicting the core values of democracy. No doubt, ignoring the challenges posed by climate change has been detrimental to democracies and will continue to have in the future.

HELP IS NOT COMING FOR THE GLOBAL NORTH

Undoubtedly, some countries are better equipped to plan and implement strategies to deal with the challenges of climate change while also protecting democratic rights. Despite some formal commitments, such as submitting plans to reduce emissions as per the Paris Agreement, there are serious doubts over the world's seriousness. Particularly the countries most at risk of being severely affected by climate change need to do their utmost to survive this climate crisis. In these countries, adaptation measures are mainly insufficient and democratic systems are most at risk. Many of the countries most at risk are being ruled by authoritarian governments or struggling democracies under the spell of populism. The primary aim of these ruling regimes has been to be in power at any cost and not to plan and invest in the long term. Therefore, measures to mitigate and adapt to climate change have not been high on the agenda. Even in some democracies, the election cycle forces parties to think of short-term glories rather than implementing long-term goals.

While poorer countries with developing economies to a large extent blame the North for climate change, they also risk being most affected by their own inaction. This will come at a cost which can cause amplified political and economic crises. The North-South divide in the climate change debate has existed since the first talks of protecting the environment in international forums. Already, at the first international conference on environmental issues in 1972, the Stockholm UN Conference on the Human Environment, the requested emission cuts were perceived to be unfair. Mainly, by many countries in the Global South, as the countries in the Global North already had developed strong economies thanks to the industrial revolution. If the less developed economies also had to cut their emissions, their possibilities to develop their economies would be constrained. Therefore, some countries perceived it as an attempt by the Global North to retain its global power position (Uddin, 2017). As the countries in the Global South strive for economic development, they face a dilemma: reducing carbon emissions often means sacrificing development goals, while pursuing development can lead to increased emissions. This no-win situation creates a complex dynamic where their efforts to protect the environment could hinder their progress towards economic growth and improved living standards. In other words, these countries are caught in a paradoxical struggle, seeking to balance sustainable development with environmental conservation, often with limited resources and technological support.

The recent global agreements are reflecting the historical accountability of the Global North and partially taking the onus. Efforts to reduce emissions and adapt to climate change can be costly. Countries therefore risk the possibility of slow economic growth. In fact, the continued North-South tensions regarding climate change mitigation and adaptation efforts still mainly concern the conflicting interests of environmental protection and economic development. Global South urges Global North to reduce emissions and help to pay for and implement efforts to limit the consequences of climate change. On the other hand, the emissions per capita in the European Union and the United States are decreasing, while the emissions per capita in countries with developing economies are increasing, except for in sub-Saharan Africa. Therefore, developed economies sometimes use this argument to blame countries in the Global South for not taking their part in the efforts to limit global warming (Vigna and Friedrich, 2023).

The ongoing blame game hinders real change in arresting the climate crisis. Although Global North is historically responsible for climate change and even has somewhat agreed to compensate for the damages they have caused, current efforts showcase that they will not pay up as much and fast as required. For instance, although Global North has set up funds to help pay for the damages caused by climate change, they have so far failed. The goal set in 2009 to mobilise US$100 billion annually by 2020 as climate finance support for climate risk countries has not yet been achieved. Also, the Loss-and-Damage Fund created in the COP27 in Sharm El-Sheikh has yet to show how it is going to be implemented. The world is reluctant to commit to US$100 billion annually to save the planet, whose survival is under serious threat from climate change. However, the world has spent more than US$2.2 trillion on the military in 2022 alone (SIPRI, 2023a). To put the amount of US$100 billion in another perspective, only one entrepreneur, Elon Musk personal wealth, has increased US$96.6 billion in just six months, from US$755.4 to US$852 billion from January to June 2023 (Massa et al., 2023). In other words, it is not the money for climate finance in short supply, but the willingness and commitment of the rich nations are equally, if not more, important.

Countries in the Global North should be held accountable and take responsibility for the unjust consequences of climate change they have created. However, as the Global North is not leading the path adequately, countries in the Global South cannot stand on the sideline and wait for the help they may never receive, while their populations are suffering. It might seem unfair, but the inaction of the leaders in the most climate-sensitive countries is a major threat to their populations. Blaming Global North for its previous and current actions and inactions will not solve the problem for the most vulnerable populations that already face the disastrous effects of climate change, and risk being even more affected in the future. There is no gain in blaming the disease if you refuse to take medicine and suffer. Climate change will not read history books to limit its impact on the countries that have caused it. For their survival and well-being, all countries must do what they can and what they should do. Given the fast-worsening climate crisis, all countries must prioritise taking adaptation measures to cope with these challenges, regardless of where on the planet it is based or their responsibility to the climate crisis.

Currently, the adaptation measures in the Global South are mainly inadequate. The International Institute for Sustainable Development has carefully reviewed the adaptation actions in 15 Asian and African countries. They concluded that the progress of climate adaptation varies between the countries. However, it had less to do with their economic strength, rather than prioritising the issues. In fact, countries with higher economic development were found to be less actively engaged in climate adaptation planning and actions compared to countries with lower economic development (Parry and Terton, 2016). Therefore, it is not merely financial power or climate risks that steer the adaptation planning. Instead, political will and leadership play an important role in the country's climate efforts.

However, in the process of adapting to climate change and reducing emissions, countries should not only have a national-level blueprint for adaptation and mitigation just to tick the box. Instead, it should also contain a detailed strategy for its implementation. It is also important for countries that are most at risk of being severely affected by the changing climate to come together and create a strong block to pressurise the major emitters. One country or a small group of countries will be unable to solve the climate crisis on their own. Additionally, the adaptation strategy of one country or some countries can have negative effects on other country or countries. For instance, many national water management projects in the Nile Basin focus on meeting global climate change but these projects adversely affect the basin's short-term and long-term water security. These projects often take national concerns into account, rather than regional ones and this can also negatively affect international efforts to prepare for climate change. Moreover, it is not uncommon that in the name of reducing carbon emissions and increasing renewable energy production, large dams are being built ignoring their contributions to global warming. In other words, the climate change debate can be used to subvert the climate action plans on the transition to green energy. In addition, in countries with ongoing conflicts or political instability, climate change tends to be pushed down as a priority concern compared to the other immediate social and economic problems experienced in society (Earle et al., 2015).

As explored earlier, a combination of misinformation and a lack of political will has kept the world from taking concrete and decisive steps towards climate action. Political leaders need to make tough decisions that carry risk and political costs, and the reality is that despite unequivocal evidence political leaders are engaged in climate politicking rather than leadership. Leaders in the Global North talk grand ideas, but most try everything they can to avoid committing to anything substantial. On the other hand, leaders of some less economically developed countries in the Global South, instead of preparing to adapt to a changing planet, are taking the cover of history and spending valuable time and energy blaming the Global North. No one can deny the role of rich and industrialised countries in contributing most to creating this climate crisis. However, simply emphasising this fact won't help anyone escape the inevitable.

Democratic leaders must prioritise decisive action to mitigate climate change while implementing policies that help their citizens adapt to climate-related challenges. This involves transitioning to sustainable energy sources, investing in climate-resilient infrastructure, and implementing regulations to reduce greenhouse gas emissions. At the same time, democracies must prioritise social justice by ensuring that climate policies benefit all citizens, particularly those most vulnerable to the impacts of climate change.

SUMMING-UP

This chapter has delved into the intricate interplay between climate change, economic development and political governance, uncovering the ethical dimensions that underpin these global challenges. The disparity between the Global North and South is not just a matter of resources or geography; it is deeply rooted in historical, ethical considerations. The ethical imperative in addressing climate change lies in acknowledging this disparity and acting to rectify it. The reluctance of the Global North to provide sufficient support and the focus of vulnerable countries on immediate political survival rather than long-term climate strategies highlight a significant ethical dilemma. The issue transcends mere environmental concern, touching upon the moral responsibility of both developed and developing nations to not only recognise their roles in the climate crisis but to actively engage in equitable solutions.

As we have seen, the path to effective climate action is fraught with challenges, including misinformation, political manoeuvring and a lack of commitment to sustainable practices. However, the ethical dimension of climate change demands a re-evaluation of priorities. It calls for a move beyond short-term political gains and economic growth, urging a focus on long-term sustainability, equity and the well-being of all global citizens. The responsibility to act ethically in the face of climate change is not limited to any one nation or group of nations. It is a collective moral obligation that spans across borders, cultures and economies. The Global South, despite its historical lack of contribution to the crisis, cannot afford to remain passive, just as the Global North must step up its efforts in providing meaningful support and leadership. The path forward involves a collaborative, international effort that not only addresses the immediate impacts of climate change but also works towards a more just and sustainable future for all.

This chapter, therefore, serves as a call to action, urging leaders worldwide to embrace the ethical imperatives of climate change. It challenges them to implement policies and strategies that not only mitigate the effects of climate change but also promote social justice and equity. By prioritising sustainable energy, climate-resilient infrastructure and inclusive policymaking, we can hope to address the ethical dilemmas posed by climate change and move towards a more equitable and sustainable world. To do that successfully, the leaders must put climate change as a national security issue in the public discourse so they possibly can get support from their electorates. This will help the leaders to do what they should do while not paying for it politically.

8
CONCLUSION

Climate change is no longer a distant environmental concern but a pressing security threat impacting global stability. This book has emphasised at the outset the transformation of security concepts over time and the emergence of climate change as a critical military as well as non-military threat. Traditional security once focused only on military might and guarding borders, with the Cold War era highlighting these concerns. However, the post-Cold War period witnessed a significant shift, recognising a spectrum of challenges beyond military threats, including civil wars, terrorism, cyber and hybrid wars and infectious diseases. The discourse on security now balances between traditional military concerns and a broader perspective that includes human security. Introduced by the United Nations Development Programme in 1994, human security shifts focus from armament to sustainable human development, emphasising the protection of human rights and safety. This broader perspective acknowledges the interconnected nature of global challenges and their cross-border impacts.

In this context, climate change emerges as a profound comprehensive threat, reshaping the concept of climate security. Human activities since the Industrial Revolution have caused a rapid rise in global temperatures, leading to various environmental and societal consequences such as heatwaves, wildfires, changes in weather patterns, droughts, flooding, water and food security issues, damage to homes and livelihoods and population displacement. This uneven impact triggers large-scale migration, insecurity in water-sharing arrangements, fuels tensions and conflicts, and affects economic activities, making climate change a 'threat multiplier' that exacerbates vulnerabilities and heightens conflict risks. Climate security has expanded the traditional understanding of security to encompass environmental issues. International and national security policies are gradually integrating climate considerations. Non-state actors like non-governmental organisations (NGOs), academia and the private sector also play crucial roles in advancing the climate security agenda. Despite international efforts, greenhouse gas emissions and temperatures continue to rise, emphasising the urgent need for nation-states to enforce decisive climate policies and ensure equitable solutions.

In this book, we delve into a comprehensive analysis of the seriousness of the climate crisis and examine how political leadership can garner popular acceptance to enact necessary measures. At both national and international levels, political leaders often demonstrate a troubling apathy towards addressing the climate crisis. This apathy is characterised by short-term thinking and compromises that diminish the urgency of environmental threats. Moreover, protectionist stances in international negotiations frequently prioritise immediate national interests over the imperative of long-term global survival. The pervasive influence of misinformation and lobbying by vested interests further clouds public perception and impedes progress in combating climate change. Despite notable agreements like the Paris Agreement, the ongoing rise in temperatures and emissions underscores the inadequacy of current political responses.

The chapter on the military examines how military operations constitute a significant contributor to global greenhouse gas emissions and environmental degradation. Paradoxically, the military itself is increasingly vulnerable to the impacts of climate change, which disrupt their operations and necessitate an expansion of their role in disaster relief efforts. However, endeavours to curtail military emissions often clash with operational requirements, illustrating the complex challenge of reconciling national security imperatives with environmental preservation.

The chapter on geopolitical shifts discusses how the phenomenon of climate change is not confined to environmental effects; it is also reshaping geopolitical landscapes. Alterations in river courses, the retreat of glaciers, melting Arctic ice and rising sea levels are redrawing international borders and intensifying competition over natural resources. These shifts lead to territorial disputes, geopolitical instability and heightened concerns regarding national security. Disputes over vital resources such as rivers (e.g. the Mekong), glacial melting in regions like the Alps and the Himalayas and the threat of rising sea levels imperilling the sovereignty of small island nations illustrate the urgency of a coordinated global response. Such a response must involve the adaptation of legal frameworks and the fostering of international cooperation to address these multifaceted challenges.

Another chapter describes how the critical aspect of the climate crisis is its impact on water security. Changes in the water cycle result in water scarcity, floods, pollution and threats to hydro-projects, culminating in a global water crisis that exacerbates political tensions and poses the risk of armed conflicts. While shared water resources often lead to conflicts, they also present opportunities for cooperation, as evidenced by successful management efforts in rivers such as the Rhine, Mekong and Ganges. Nevertheless, current global frameworks inadequately address the water crisis aggravated by climate change, necessitating a flexible and adaptive approach to water management alongside enhanced global cooperation.

The chapter on migration shows how the issue of climate-induced migration is rapidly evolving from a peripheral concern to a major international issue. The resurgence of nationalism and populism complicates the response to this crisis, potentially fuelling tensions and conflicts. Moreover, the lack of legal recognition and protection for climate migrants under international law poses a significant challenge. Addressing this issue requires a substantial re-evaluation of international policies and legal frameworks to ensure the protection and accommodation of climate migrants, underscoring the urgent need for a comprehensive and coordinated global response.

Finally, this book delves into the ethical dimensions of climate change, highlighting the glaring disparity between the contributions of the Global North to greenhouse gas emissions and its insufficient actions compared to the struggles of the Global South in confronting environmental upheavals. Recognising and rectifying this ethical imperative demands a collaborative, international effort that transcends borders, cultures and economies. Such an effort must prioritise addressing the immediate impacts of climate change while striving towards a more equitable, just and sustainable future for all inhabitants of our planet.

The climate crisis, in many respects, has transcended its traditional boundaries as an environmental issue, asserting itself as an urgent national security concern that threatens the stability and safety of nations worldwide. The impacts of climate change – extreme weather events, rising sea levels and shifting weather patterns – are not just ecological phenomena; they directly affect the security and economic stability of countries. Climate change exacerbates existing vulnerabilities, from straining resources to igniting geopolitical tensions. For example, the scarcity of water and arable land due to changing climatic conditions can lead to intensified competition among nations and communities, increasing the risk of conflict. Additionally, natural disasters, amplified in frequency and intensity by climate change, challenge the readiness and resilience of national defence systems, stretching military resources thin as they increasingly engage in disaster response and humanitarian aid, alongside their traditional defence roles.

The urgency of addressing the climate crisis as a national security issue is further underscored by its global scope and the interconnected nature of modern geopolitics. The effects of climate change know no borders, making international cooperation crucial. Rising sea levels and extreme weather events can trigger mass migrations, leading to regional instability and creating new security challenges for countries. These climate-induced migrations, coupled with the loss of habitable land, can potentially lead to increased political tensions and even conflict, as nations grapple with the influx of displaced populations and the resultant strain on resources. Moreover, the climate crisis poses significant risks to critical infrastructure, from energy grids to transportation networks, which are essential for national security. The failure to address these risks could lead to widespread disruption and have cascading effects on a country's security, economy and societal well-being. Therefore, mitigating the effects of climate change is not only an environmental imperative but also a strategic necessity for national and global security, requiring coordinated efforts at both national and international levels.

The preceding chapters highlight the urgent need for a global, cooperative response to the multifaceted challenges posed by climate change. It underscores the importance of understanding and addressing the broad spectrum of climate-related security issues, from traditional military concerns to emerging geopolitical, environmental and human security challenges. The call to action is clear: global leaders must prioritise sustainable, equitable climate policies and strategies to ensure a secure and sustainable future. It is not anything new that the political leaders are not aware of it. They have been attending the world climate conferences for almost three decades now. The alarming climate trends we are observing are impossible to overlook. The world is experiencing a full-blown climate crisis – with extreme heat waves, devastating storms and melting glaciers. So, why political leadership of the world is not doing what it is supposed to do? As we notice, climate negotiations and governance have become a crowded field. The last UN Climate Conference COP28 in the United Arab Emirates witnessed a record number of delegates, a testament to the escalating global concern and engagement on climate issues (McSweeney, 2023). However, this surge in attendance, particularly the involvement of a diverse array of non-state actors while there is a climate treaty in place since 2015, necessitates a critical examination of the efficacy of this 'inclusive' climate governance.

NATION-STATE

The climate crisis represents one of the most daunting and urgent challenges facing humanity today. It is characterised by a series of interlinked phenomena, primarily driven by the unprecedented levels of greenhouse gases emitted by human activities, leading to global warming. The manifestations of this crisis are diverse and far-reaching, including rising global temperatures, melting ice caps and glaciers, more frequent and severe weather events like hurricanes, droughts and floods and rising sea levels. These changes are having profound impacts on natural ecosystems, wildlife and human societies. They disrupt agricultural productivity, threaten water resources and lead to the displacement of populations, thereby exacerbating issues like poverty, food insecurity and social inequality. The crisis calls for immediate and concerted action. Mitigation efforts like reducing carbon emissions, transitioning to renewable energy sources, and conservation initiatives, coupled with adaptation strategies such as building resilient infrastructure and developing sustainable agricultural practices, are essential to address the multifaceted challenges posed by the climate crisis.

Historically, climate negotiations and governance were predominantly the realm of nation-states. The recent COPs, including the staggering 97,000 delegates at COP28, reveal a dramatic shift. Of these, only 24,488 represent the nation-states – the primary parties to the convention (McSweeney, 2023). The rest include a broad spectrum of non-state actors: environmental NGOs, activist groups, intergovernmental organisations, city networks, businesses, indigenous communities and others. This trend spurred post-Copenhagen Conference in 2009, signifies a more dispersed, polycentric approach to climate governance.

While the inclusion of non-state actors democratises the negotiation process, it also introduces complexities. These actors bring divergent, often conflicting agendas. Some businesses push eco-friendly policies; others indulge in greenwashing or advocate deregulation. NGOs range from those seeking insider status to those demanding radical systemic changes. The rise of the climate justice movement further diversifies this landscape with its energetic activism and global protests. Though the states can hold some closed-door meetings, however, the myriad of voices at the climate summits, while enriching the dialogue, complicates the negotiation process. Each participant, wielding different levels of power and influence, pursues distinct goals and strategies. The ensuing interplay of these diverse agendas and power structures elongates and complicates the negotiation process. It shifts the focus away from a unified strategy, leading to diffusion and dilution of state-level efforts. Moreover, this multifaceted approach often results in a blame game, where responsibility for climate action becomes fragmented and elusive.

At COP28 in Dubai, for instance, US Climate Envoy John Kerry's criticism of US oil producers for not doing enough to reduce the effects of the climate crisis exemplifies this trend (Dlouhy and Fraher, 2023). Such statements, while spotlighting corporate responsibilities, inadvertently deflect from the fundamental role of the state in enacting and enforcing environmental laws and regulations. The state's unique position in this arena is defined by its sovereign authority, legal power and enforcement capacity. Unlike non-state actors, states can enact laws and policies that address climate change challenges effectively and comprehensively.

This legislative power is complemented by the state's ability to enforce these laws through various mechanisms, including regulatory agencies, environmental protection bodies and judicial systems. Moreover, states can integrate climate concerns into broader policy areas such as urban planning, energy policy, transportation and industrial regulation. States possess the financial resources and institutional infrastructure necessary for large-scale initiatives on climate change mitigation and adaptation. Their ability to mobilise funds, enforce regulations and incentivise sustainable practices is unparalleled. Governments, with their ability to mobilise significant financial resources, are uniquely positioned to fund large-scale climate projects and research, which are often beyond the reach of private entities or NGOs.

The battle against climate change is also inherently global, transcending national borders and necessitating international collaboration. In other words, international climate agreements, pivotal in the global response to climate change, rely on state ratification and implementation. When a state ratifies an international climate treaty, it signifies a commitment to adhere to the treaty's provisions and integrate them into national law. This process is vital because it translates global objectives into national action, making the global fight against climate change tangible and actionable at the local level.

No doubt, the battle against climate change requires a multifaceted approach. While non-state actors contribute innovation, grassroots mobilisation and pilot new ideas, the state is the conductor of this symphony. It sets the tempo and ensures a harmonised collective effort towards common goals.

In the urgency of the climate crisis, the involvement of an excessive number of non-state actors at negotiation tables is therefore becoming counterproductive. It is making climate governance more cumbersome and detracts from the efficiency and focus required for immediate and effective decision-making. The adage 'too many cooks spoil the broth' is apt in this context. The sheer number of participants and the diversity of their agendas dilute the focus and efficacy of climate governance, leading to delayed suboptimal outcomes.

To effectively address the climate crisis, particularly in achieving global net-zero emissions and securing necessary climate financing, it's crucial to reaffirm and prioritise the primary role of nation-states in negotiations. For cohesive and decisive action, as well as the formation of binding international climate agreements, the main responsibility must rest with the nation-states. They possess the legal authority, financial resources, enforcement capabilities and diplomatic influence crucial for impactful global climate action. Additionally, they bear the responsibility to safeguard the rights and interests of their historically, socially and economically disadvantaged citizens during climate negotiations.

Nation-states are pivotal in enacting laws, mobilising resources and establishing international treaties, all of which are essential for a coordinated and effective global response to climate change. They have historically signed and administered several international treaties, such as the UN Convention on the Law of the Sea, the Antarctic Treaty, the Outer Space Treaty, the Montreal Protocol on Substances that Deplete the Ozone Layer and the Convention on Biological Diversity, to protect global commons. Given this track record, why shouldn't nation-states be entrusted with the responsibility to protect the planet from climate change? While the involvement of various actors in climate negotiations reflects global concern and commitment to addressing climate change, it is vital to recognise and reinforce the central role of nation-states.

NATIONAL SECURITY

Picture a world perched on the edge of a precipice, facing the formidable adversary of the climate crisis. This critical battle transcends the efforts of philanthropists, NGOs and youth movements, and how valiant they might be. While their contributions are crucial, the real vanguards in this existential struggle are nation-states. Visualise political leaders across the globe, endowed with the power to enact monumental changes. The question looms: are they rising to the challenge? The destiny of our planet hinges on their commitment to combat global warming, rising sea levels and the increasing frequency of natural disasters.

The looming spectre of climate change casts a vast, urgent shadow and its repercussions are felt worldwide. Our planet already grapples with the brunt of this crisis, a situation only expected to intensify. Citizens worldwide are raising their voices, demanding decisive action from their governments, far beyond empty promises. A burgeoning climate movement echoes a resounding call for tangible steps. Even schoolchildren, our future custodians, are globally mobilising with a straightforward yet profound plea to the powers: 'Act responsibly, protect our future'.

The critical obstacle lies in a conspicuous lack of political will. This inertia has restrained the world from taking the bold, necessary measures to address climate change. Political leaders are at a juncture where hard choices come with risks and political consequences. However, the undeniable truth is that the long-term repercussions of unchecked global warming far surpass the immediate sacrifices needed for climate action. Leaders, often confined by the practicalities of politics and election cycles, now face the urgent need for strong, equitable and wise leadership, acknowledging the extensive benefits of collaborative international efforts.

Despite undeniable evidence of climate change's devastating impacts, world leaders continue to downplay its urgency. Many either neglect the crisis or get entangled in blame games. The political class's lack of commitment to addressing climate change has led to the repeated failures of important climate conferences and negotiations. It has become almost clichéd for world leaders to vow to stop climate change for future generations without addressing the immediate, pressing need for a global response. There's a widespread, erroneous belief that climate change is a future issue, not a present reality.

As we have already discussed in the book, globally, temperatures have risen by more than 1°C since the last century, with the rate of warming having doubled in the last 40 years. This uneven warming pattern has disproportionately affected certain regions. For instance, the Arctic is warming faster than the rest of the planet. In the Middle East, which is warming at twice the global average, extreme temperatures and droughts have severely impacted economic activities and agriculture. Recent extreme weather events have shattered historical records. The World Meteorological Association reports a fivefold increase in climate and weather-related disasters over 50 years, leading to massive economic losses. Political leaders and the media often speak of climate change-induced migration as a future possibility, but it is already happening. Sea levels have risen significantly, altering settlement patterns and socio-economic conditions in coastal and island regions.

Climate change is already influencing water supply and demand, leading to increased disputes over shared rivers. It also plays a significant role in causing civil wars. Conflicts in regions like Sudan, Syria and the Lake Chad basin have been partially attributed to climate change, which undermines livelihoods and lives. Military coups are experiencing a resurgence in countries most affected by climate risks, further destabilising these regions.

In other words, the impacts of climate change are not mere future concerns; they are current realities, spawning vast ecological, economic, social, political and humanitarian crises. World leaders must prioritise climate change and take immediate action. Echoing the words of Mary Robinson, Chair of the Elders, during the COP26 Glasgow negotiations, 'Now is the moment for decisive action, not obfuscation or half measures' (Robinson, 2021). The era of rhetoric has passed; the era of action has arrived. The consequences of inaction are tangible, widespread and catastrophic. Our planet, our shared home, demands an immediate, united effort to mitigate the impacts of climate change. The future is unfolding now, calling for our urgent and undivided attention.

At this pivotal moment, as the clock ticks inexorably towards a point of no return, humanity's fate rests on the shoulders of political will and leadership. We need visionary leaders to unite their people in a collective journey towards a sustainable future. This is a call for bold leadership, a collective leap into the unknown, armed with a determination to protect our planet. Hesitation must give way to decisive action. However, a glaring disconnect exists between the stirring speeches and photo-op-filled annual climate summits, and the actual implementation of change. Most leaders are mired in the blame game, pointing fingers for climate woes rather than taking substantive action. There is a perplexing mismatch between the public perception of climate threats and the priorities of political leaders. The reluctance to frame climate change as a national security threat, for fear of it leading to increased military spending and isolationist policies, has inadvertently made it a secondary issue in elections and policymaking.

Climate change is a 'threat multiplier,' shaking the foundations of national peace, security and stability. To elevate this issue in policy agendas, a seismic shift in dialogue is needed. It's time to highlight how climate change is eroding national security. Consider this: rising temperatures are impacting military missions and degrading military hardware. Soldiers face new health challenges in changing climates. Natural disasters like floods and hurricanes are diverting military resources to rescue and relief efforts. Rising sea levels and melting glaciers are redrawing national borders, inciting disputes over economic zones. Climate change could trigger conflicts over shared rivers and drive movements of climate refugees, intensifying international tensions.

Additionally, climate-induced volatility in food prices and economic downturns undermine social cohesion and government legitimacy, potentially leading to political instability, rebellions and even military coups. Emphasising climate change's impact on national security doesn't detract from its human security threat. Instead, it strategically captures political leaders' attention, urging them to take decisive, urgent measures against global warming. The message is clear: unless addressing climate change becomes a top priority for political leaders, individual, business and civil society efforts, commendable as they are, will simply be insufficient. This is a clarion call to arms, a rallying cry for world leaders to recognise the urgency of the climate crisis and take swift action to save our planet. The time for half-hearted measures and political posturing has ended. The future of our world depends on bold, transformative actions taken now.

For decades the attempts to frame climate crisis under the frame of human security have neither resulted in halting climate change nor making any satisfactory progress in helping the affected people to adapt. The world continues to struggle to effectively address the human security crisis exacerbated by climate change. Despite the commitments made under the Paris Agreement, the world remains off track in limiting global warming. The energy security concerns and disputes over compensation for climate-related damages further complicate emission reduction agreements. However, financial challenges remain the central obstacle in climate governance. The disparity in contributions and commitments from developed nations, the underfinancing of climate action and the failure to meet the promised financial support to developing nations highlight a significant shortfall in climate finance. This shortfall undermines trust in climate talks and impedes efforts to mitigate the worst impacts of climate change. Adaptation is now critical for survival, but progress is extremely slow. Securing sufficient contributions remains complex, as developed countries have to balance supporting climate-risk nations with their own domestic political and economic agendas. At COP15 in Copenhagen in 2009, wealthy countries promised to provide US$100 billion annually to developing countries for climate adaptation and mitigation. In 2020, the total climate finance reported was US$83.3 billion, increasing to US$89.6 billion in 2021. However, Oxfam estimates the actual value of financial support for climate action to be only between US$21 and US$24.5 billion, citing misleading accounting practices and a lack of adequate grants as exacerbating factors (Oxfam, 2023). The United Nations Environmental Programme in its recent report 'Underfinanced Underprepared' declares the adaptation gap in climate financing a critical emergency (UNEP, 2023b). Despite the establishment of a 'Loss-and-Damage Fund', adaptation funding has decreased in recent years. It will be safe to conclude that the world has primarily failed to adequately address human security in the face of the climate crisis. So, why not reframe the climate crisis as a national security issue?

Framing an issue as a matter of national security often facilitates bold decision-making by political leaders for several reasons. Firstly, national security concerns typically command high levels of public attention and urgency. The perception that a nation's safety and well-being are at stake can create a powerful impetus for action, rallying public support behind the government. This broad-based support can provide political leaders with a stronger mandate to implement decisive measures that might be more challenging under normal circumstances. The heightened sense of urgency and the perceived need to protect national interests can also lead to a more unified stance among political factions, reducing the usual partisan divides and facilitating quicker decision-making.

Moreover, national security issues often provide political leaders with greater latitude in their decision-making. In many countries, laws and constitutional provisions grant the government special powers in times of national crisis, allowing for swift and extraordinary measures. This can include bypassing the usual bureaucratic and legislative processes, which can be time-consuming and fraught with obstacles. The framing of an issue as a national security concern can also lead to increased secrecy and less public scrutiny, as the details of the government's response may be classified for security reasons. This combination of heightened authority reduced bureaucratic hurdles, and decreased public scrutiny creates an environment in which bold and decisive actions can be taken more easily by political leaders.

The media often gives extensive coverage to national security issues, which can help leaders in shaping public opinion and rallying support for their policies. Historically, leaders who have successfully navigated national security challenges are often viewed favourably in the annals of history. This precedent can motivate current leaders to take bold actions in similar situations. National security issues often involve international relations and diplomacy, areas where heads of state and government leaders typically have more direct control. This allows them to take decisive actions that might be more challenging in domestic policy areas.

Undoubtedly, the framing of the climate crisis as a matter of national security can create a conducive environment for political leaders to take bold decisions, supported by public consensus, expanded powers and a focus on the urgent need to protect the people. The climate crisis, once recognised as a national security threat, not only gains the urgency it demands but also opens pathways for more assertive, immediate and effective action by political leaders. The future is unfolding now, and it calls for our urgent, unified and resolute response.

The framing of the climate crisis in terms of national security is a strategic approach that can enable political leaders to take more urgent and decisive actions to address both the mitigation and adaptation aspects of this global challenge. This perspective emphasises the direct and indirect threats that climate change poses to a nation's stability, sovereignty and safety, thereby making it a priority on the national agenda. Framing the climate crisis in terms of national security enables a comprehensive approach to both mitigation and adaptation. Mitigation efforts, such as reducing greenhouse gas emissions and transitioning to renewable energy sources, become matters of national strategic importance. Similarly, adaptation strategies, including bolstering infrastructure against extreme weather events and developing resilient food and water systems, are seen as essential to maintaining national security. This framing provides a compelling rationale for undertaking large-scale, transformative actions that might otherwise be politically difficult to justify.

In addition, considering the climate crisis as a matter of national security can lead to more effective international collaborations. Nations often come together more readily to address common security threats. This collective approach can lead to the sharing of resources, knowledge and strategies, making global efforts to combat climate change more coordinated and impactful. It also places climate change in international security discussions, elevating its priority in global policy agendas. However, political leaders must balance this national security framing with the recognition that climate change is a global issue requiring a unified global response. While national security concerns can catalyse action, they must not lead to isolationist policies or undermine international cooperation. At the same time, they must also take into account the diverse voices of development, environment and climate issues within their own countries. The global nature of the climate crisis necessitates a shared responsibility and collective action, transcending national borders and interests. In other words, while national security framing is a powerful tool for mobilising action, it should be complemented with efforts to foster global solidarity and cooperation in facing this unprecedented challenge.

In conclusion, the urgency of addressing climate change cannot be overstated. As outlined in this book, the climate crisis presents multifaceted challenges that require immediate and concerted action from political leaders worldwide. To effectively combat climate change, countries must prioritise it as a matter of national security. This entails recognising the interconnectedness of environmental, social and economic stability and making necessary compromises to cooperate with the rest of the world. By acknowledging climate change as a national security issue, countries can transcend short-term interests and prioritise long-term global survival. This paradigm shift requires political courage and foresight to implement policies that mitigate greenhouse gas emissions, adapt to environmental changes and foster international cooperation. Furthermore, it necessitates re-evaluating traditional security frameworks to encompass the complex challenges posed by climate change, including resource competition, migration pressures and geopolitical instability.

To succeed in this endeavour, countries must transcend partisan divides and ideological differences, recognising that the consequences of inaction far outweigh any perceived short-term benefits. By working together, nations can leverage their collective resources, expertise and influence to develop innovative solutions and enact meaningful change. This entails investing in renewable energy in a humane manner, implementing sustainable land and water management practices and fostering resilience in vulnerable communities while prioritising their interests. Ultimately, the path forward requires a fundamental shift in mindset – from viewing climate change as a distant threat to understanding it as an existential challenge that demands immediate and decisive action. By prioritising climate change as a national security imperative, countries can forge a more sustainable and secure future for generations to come. The time for complacency is over; the time for bold and decisive leadership is now.

REFERENCES

Acaps. 2023. *Horn of Africa: Impact of Drought on Children*. Thematic report. 24 April 2023.
Afifi, T. and Jäger, J. 2011. *Environment, Forced Migration, and Social Vulnerability*. Berlin: Springer.
Aftenposten. 2021. *Norske velgere har endret mening. To saker er blitt mye viktigere i løpet av sommeren*. 19 August 2021.
Ahlbom Persson, T. and Povitkina, M. 2017. 'Gimme Shelter': The Role of Democracy and Institutional Quality in Disaster Preparedness. *Political Research Quarterly*. Vol. 70. No. 4.
Anisimov, A. and Magnan, A. K. 2023. *The Global Transboundary Climate Risk Report*. The Institute for Sustainable Development and International Relations & Adaptation Without Borders. 114 pages.
APA (American Psychiatric Association). 2023. *Climate Change and Mental Health Connections*. [online] Available at: https://www.psychiatry.org/patients-families/climate-change-and-mental-health-connections [2023-10-02].
Arctic Council. 1996. *Declaration on the Establishment of the Arctic Council (The Ottawa Declaration)*.
Arctic Review. 2022. *Northern Sea Route*. [online] Available at: https://arctic.review/future/northern-sea-route/ [2023-05-30].
Arezki, R. and Bruckner, M. 2011. *Food Prices and Political Instability*. Working Paper No. 2011/062. International Monetary Fund. 1 March 2011.
Atherton, K. 2021. *U.S. Forces Are Leaving a Toxic Environmental Legacy in Afghanistan*. Scientific American. August 30 2021.
Barquet, K. 2015. 'Yes to Peace'? Environmental Peacemaking and Transboundary Conservation in Central America. *Geoforum*. Vol. 63. Pp. 14-24.
BBC. 2014. *Denmark Challenges Russia and Canada over North Pole*. 15 December 2014.
BBC. 2020. *US Election 2020: Biden Seeks to Clarify Remark on Ending Oil*. 24 October 2020.
Behrman, S. and Kent, A. 2020. The Teitiota Case and the Limitations of the Human Rights Framework. *Questions of International Law*. 30 November 2020.
Behrman, S. and Kent, A. 2022. *Climate Refugees: Global, Local and Critical Approaches*. Cambridge: Cambridge University Press.
Bell, J. et al. 2021. Pathways to Sustaining Tuna-dependent Pacific Island Economies during Climate Change. *Nature Sustainability*, Vol. 4. No. 10. Pp. 900-910.
Bernauer, T. and Böhmelt, T. 2020. International Conflict and Cooperation over Freshwater Resources, *Nature Sustainability*, Vol. 3. No. 5. Pp. 350-356.
Bettini, G., Beuret, N. and Turhan, E. 2021. On the Frontlines of Fear: Migration and Climate Change in the Local Context of Sardinia, Italy. *ACME: An International Journal* for Critical Geographies, 20(3), 322-340. https://doi.org/10.14288/acme.v20i3.1838

Bettini, G. Nash, S.L. and Gioli, G. 2017. One Step Forward, Two Steps Back? the Fading Contours of (In)justice in Competing Discourses on Climate Migration. *The Geographical Journal*, Vol. 183. Pp. 348-358.
Bilal, A. 2021. 'Hybrid Warfare – New Threats, Complexity, and "Trust" as the Antidote', *NATO Review*. 30 November.
Billing, L. 2023. America's War in Afghanistan Devastated the Country's Environment in Ways that May Never Be Cleaned up. *Inside Climate News*. 25 September 2023.
Birnbaum, M. 2022. As Wildfires Grow, Militaries Are Torn between Combat, Climate Change. *The Washington Post*. 26 September 2022.
Bogren, J., Gustavsson, T. and Loman, G. 2008. *Klimat och väder*. Lund: Studentlitteratur.
Borgomeo, E., Jägerskog, A. Zaveri, E. Russ, J. Khan, A. and Damania, R. 2021. *Ebb and Flow, Volume 2: Water in the Shadow of Conflict in the Middle East and North Africa*. Washington, DC: World Bank.
Brown, M. and Bohrer, B. 2023. Alaska Oil Project Approval Adds yet Another Climate Concern. *AP News*. 15 March 2023.
Busby, J. 2022. *States and Nature: The Effects of Climate Change on Security*. Cambridge: Cambridge University Press.
Bushard, B. 2022. Record Flooding: $40 Billion of Damage in Pakistan as Monsoons Devastate South Asia. *Forbes*. 19 October 2022.
Buzan, B. 1983. *People, States, and Fear: The National Security Problem in International Relations*. Brighton: Harvester Wheatsheaf.
Buzan, B. 1984. Peace, Power, and Security: Contending Concepts in the Study of International Relations. *Journal of Peace Research*, Vol. 21. No. 2. Pp. 109-125.
CAAD and CAN (Climate Action Against Disinformation and Conscious Advertising Network). 2022. *The Impacts of Climate Disinformation on Public Perception*. [online] Available at: https://caad.info/wp-content/uploads/2022/11/The-Impacts-of-Climate-Disinformation-on-Public-Perception.pdf [2023-08-31].
Carlin, R. Love, G. and Zechmeister, E. 2014. Natural Disasters and Democratic Legitimacy: The Public Opinion Consequences of Chile's 2010 Earthquake and Tsunami. *Political Research Quarterly*, Vol. 67. No. 1. Pp. 3-15.
CDP (Carbon Disclosure Project). 2017. *The Carbon Major Database: CDP Carbon Majors Report 2017*.
Caretta, M. Fanghella, V. Rittelmeyer, P. Srinivasan, J. Panday, P. Parajuli, J. Priya, R. Reddy, U. Seigerman, C. and Mukherji, A. 2023. Migration as Adaptation to Freshwater and Inland Hydroclimatic Changes? A Meta-Review of Existing Evidence. *Climatic Change*, Vol. 176. No. 100.
CLCS (Commission on the Limits of the Continental Shelf). 2012. *The Continental Shelf. Division for Ocean Affairs and the Law of the Sea, Office of Legal Affairs*, United Nations.
Clement, V., Riguad, K., de Sherbinin, A., Jones, B., Adamo, S., Schewe, J., Sadiq, N. and Shabahat, E. 2021. *Groundswell Part 2: Acting on Internal Climate Migration*. Washington, DC: World Bank.
Climate Action Tracker. 2023a. *India*. [online] Available at: https://climateactiontracker.org/countries/india/ [2023-08-18].

Climate Action Tracker. 2023b. *China*. [online] Available at: https://climateactiontracker.org/countries/china/policies-action/ [2023-08-17].

Climate Action Tracker. 2023c. *Russia*. [online] Available at: https://climateactiontracker.org/countries/russian-federation/ [2023-08-31].

Climate Watch. 2023. *Historical GHG Emissions*. [online] Available at: https://www.climatewatchdata.org/ghg-emissions?end_year=2021&source=GCP&start_year=1960 [2023-08-16].

Coalition for Urban Transitions. 2019. *Climate Emergency*, Urban Opportunity. 19 September 2019.

Cochrane, F. 2015. *Migration and Security in the Global Age: Diaspora Communities and Conflict*. New York: Routledge.

Cole, B. 2022. *Decreasing Reliance on Fossil Fuels to Increase Defense Capability*. Air and Space Power Center. [online] Available at: https://airpower.airforce.gov.au/blog/BP29542709 [2023-04-28].

Collins, S. 2021. The Mysterious Case of an Island that 'vanished' in the Gulf of Mexico. *Mexico News Daily*. 28 June 2021.

Connor, P. and Krogstad, J.M. 2018. *Many Worldwide Oppose More Migration – Both into and Out of Their Countries*. Pew Research Center. 10 December 2018.

Cook, B. Anchukaitis, K. Touchan, R. Meko, D. and Cook, E. 2016. Spatiotemporal Drought Variability in the Mediterranean over the Last 900 Years. *Journal of Geophysical Research: Atmospheres*, Vol. 121. No. 5.

Copernicus. 2023. *European Summer 2023: A Season of Contrasting Extremes*. [online] Available at: https://climate.copernicus.eu/european-summer-2023-season-contrasting-extremes [2023-11-28].

Cornwall, W. 2021. Europe's Deadly Floods Leave Scientists Stunned. *Science*. Vol. 373. No. 6553. Pp. 372–373.

Corporate Leaders Group. 2022. *More than 150 Business Leaders Call on EU to Strengthen Energy Security by Accelerating Green Transition*. University of Cambridge. 11 May 2022.

Cox, J., Lingbeek, J., Weisscher, S. and Kleinhans, M. 2022. Effects of Sea-Level Rise on Dredging and Dredged Estuary Morphology. *Journal of Geophysical Research Earth Surface*. Vol. 127. No. 10.

Crawford, N. 2019. *Pentagon Fuel Use, Climate Change and the Costs of War*. Watson Institute, Brown University.

Crawford, N. 2022. *The Pentagon, Climate Change and War: Charting the Rise and Fall of U.S. Military Emissions*. Cambridge, Massachusetts: The MIT Press.

CRED (Center for Research on the Epidemiology of Disasters). 2023. *2022 Disasters in Numbers*. Belgium: Emergency Event Data Base.

CRS (Congressional Research Service). 2019. *Military Installations and Sea-Level Rise*. July 26 2019.

CTBTO (Comprehensive Nuclear-Test-Ban Treaty Organization). 2011. *20th Anniversary of Closure of Semipalatinsk Test Site*.

Dams, T., van Schaik, L. and Stoetman, A. 2020. *Presence before Power: China's Arctic Strategy in Iceland and Greenland*. Clingendael Report. June 2020.

Daniel, L. 2011. Lynn: Energy Strategy Till Help Forces to Adapt for Future. *American Forces Press Service.* 15 June 2011.

Darbyshire. 2021. *Deforestation in Conflict Areas in 2020. Conflict and Environment Observatory.* [online] Available at: https://ceobs.org/assessment-of-recent-forest-loss-in-conflict-areas/ [2023-04-04].

Deemer, B., Harrisson, J., Li, S., Beaulieu, J., DelSontro, T., Barros, N., Bezerra-NEto, J., Powers, S., dos Santos, M. and Vonk, J. 2016. Greenhouse Gas Emissions from Reservoir Water Surfaces: A New Global Synthesis. *BioScience,* Vol. 66. No. 11. Pp. 949-964.

Desmidt, S., Puig, O., Detges, A., van Ackern, P. and Tondel, F. 2021. *Climate Change and Resilience in the Central Sahel.* CASCADES Policy Paper. CASCADES.

Dlouhy, J. and Fraher, J. 2023. *Kerry Calls Out Chevron as Lagging on Climate at COP28.* Bloomberg.

DN (Department of the Navy). 2014. *The United States Navy Arctic Roadmap for 2014 to 2030.* February 2014.

DN (Department of the Navy). 2022. *Climate Action 2030: Department of the Navy.* [online] Available at: https://www.navy.mil/Portals/1/Documents/Department%20of%20the%20Navy%20Climate%20Action%202030.pdf [2023-04-25].

DoD (Department of Defense). 2011. *Energy for the Warfighter: Operational Energy Strategy.* [online] Available at: https://www.acq.osd.mil/eie/Downloads/OE/Operational%20Energy%20Strategy,%20Jun%2011.pdf [2023-04-27].

DoD (Department of Defense). 2019. *Report on Effects of a Changing Climate to the Department of Defense. Office of the under Secretary of Defense for Acquisition and Sustainment.* January 2019. [online] Available at: https://media.defense.gov/2019/Jan/29/2002084200/-1/-1/1/CLIMATE-CHANGE-REPORT-2019.PDF [2023-03-23].

DoD (Department of Defense). 2021a. *Department of Defense Climate Adaptation Plan.* [online] Available at: https://www.sustainability.gov/pdfs/dod-2021-cap.pdf [2024-01-17].

DoD (Department of Defense). 2021b. *DOD, Navy Confront Climate Change Challenges in Southern Virginia.* [online] Available at: https://www.defense.gov/News/News-Stories/Article/Article/2703096/dod-navy-confront-climate-change-challenges-in-southern-virginia/ [2023-04-18].

DoD (Department of Defense). 2021c. *Department of Defense: Annual Energy Management and Resilience Report (AEMRR) Fiscal Year 2020.* [online] Available at: https://www.acq.osd.mil/eie/Downloads/IE/FY%202020%20AEMRR.pdf [2023-04-18].

DoE (U.S. Department of Energy). 2022. *U.S. and 30 Countries Commit to Release 60 Million Barrels of Oil from Strategic Reserves to Stabilize Global Energy Markets.* [1 March 2022].

Dominguez, G. 2023. Weathering the Storm: How Japan Is Factoring Climate Change into Defense Policy. *The Japan Times.* 19 March 2023.

Dommen, A. 1984. Laos in 1984: The Year of the Thai Border. *Asian Survey,* Vol. 25. No. 1.

Donaldson, J. 2009. Where Rivers and Boundaries Meet: Building the International Boundaries Database. *Water Policy,* Vol. 11. Pp. 629-644.

Donaldson, J. 2011. Paradox of the Moving Boundary: Legal Heredity of River Accretion and Avulsion. *Water Alternatives,* Vol. 4. No. 2. Pp. 155-170.

Dongare, A. 2022. Indian Army to Induct Electric Vehicles in its Fleet. *India Today.* 13 October 2022.

DoS (U.S. Department of State). 2022. *Joint Statement on Arctic Council Cooperation Following Russia's Invasion of Ukraine.* 3 March 2022.
DoS PM (Bureau of Political-Military Affairs). 2021. *U.S. Arms Transfer Increased by 2.8 Percent in FY 2020 to $175.08 Billion.* 20 January 2021.
Doshi, R. Dale-Huang, A. and Zhang, G. 2021. Northern Expedition: China's Arctic Activities and Ambitions. *Foreign Policy at Brookings.* April 2021.
Doyle, A. 2021. Islands, Rocks and Tuna: Pacific Nations Draw New Battle Lines against Rising Seas. *Reuters.* 11 March 2021.
Downing, S. 2023. Sullivan Puts Navy Secretary on Hot Seat for Dabbling in Climate Change while Ignoring Military Lethality. *Must Read Alaska.* 22 April 2023.
Dutch Ministry of Defense. 2021. *Defence Energy Transition Plan of Action.* New Energy in the Organisation.
Earle, A. Cascão, A.E. Hansson, S. Jägerskog, A. Swain, A. and Öjendal, J. 2015. *Transboundary Water Management and the Climate Change Debate.* London: Routledge.
Eckstein, G. 2017. *The International Law of Transboundary Groundwater Resources.* Oxon: Routledge.
Eckstein, D. Künzel, V. and Schäfer, L. 2021. *Global Climate Risk Index 2021.* Briefing Paper. Bonn: German Watch.
Enel and The European House Ambrosetti. 2021. *European Governance of the Energy Transition: Enabling Investments.* Available at: https://www.enelfoundation.org/content/dam/enel-foundation/topics/2021/09/EF_R21_digital_distese.pdf [2023-06-22].
Evans, S. 2021. *Analysis: Which Countries Are Historically Responsible for Climate Change?* Carbon Brief.
EU (European Union). 2021. *Eurobarometer: Climate Change.* July 2021.
EUvsDisinfo. 2023. *The Kremlin's Climate Camouflage.* 2 August 2023. [online] Available at: https://euvsdisinfo.eu/the-kremlins-climate-camouflage/ [2023-08-30].
European Commission. 2021. *Joint Communication to the European Parliament, the Council, the European Economic and Social Committee and the Committee of the Regions: A Stronger EU Engagement for a Peaceful, Sustainable and Prosperous Arctic.* Brussels, 13 October 2021.
FAO (Food and Agriculture Organization of the United Nations). 2021a. *The State of Food Security and Nutrition in the World 2021: The World Is A Critical Juncture.* Rome: FAO.
FAO (Food and Agriculture Organization of the United Nations). 2021b. *The State of the World's Land and Water Resources for Food and Agriculture: Systems at Breaking Point.* Synthesis Report 2021. Rome: FAO.
FAO (Food and Agriculture Organization of the United Nations). 2022. *The State of Food Security and Nutrition in the World 2022. Repurposing Food and Agricultural Policies to Make Healthy Diets More Affordable.* Rome: FAO.
FAO (Food and Agriculture Organization of the United Nations). 2023. *The State of Food Security and Nutrition in the World 2023: Urbanization, Agrifood Systems, Transformation and Healthy Diets across the Rural-Urban Continuum.* Rome: FAO.

Farstad, F. M. and Aasen, M. 2022. Climate Change Doesn't Win You a Climate Election: Party Competition in 2021 Norwegian General Election. *Environmental Politics*. Vol. 32. No. 4.

Fawthrop, T. 2017. *China's Myanmar Dam Hypocrite*. The Diplomat. 28 January 2017.

Ferrari, M. Pasqual, E. and Bagnato, A. 2018. *A Moving Border: Alpine Cartographies of Climate Change*. New York: Columbia University Press.

Fiorino, D. J. 2018. *Can Democracy Handle Climate Change?* Cambridge: Polity Press.

Fletcher Primer, 2023. *Law of the Sea: A Policy Primer*. The Fletcher School of Law and Diplomacy, Tufts University.

Fuchs, R., Brown, C. and Rounsevell, M. 2020. Europe's Green Deal Offshores Environmental Damage to Other Nations. *Nature*, Vol. 586. Pp. 671–674.

French Ministry of the Armed Forces. 2022. *Climate & Defense Strategy*. [online] Available at: https://www.defense.gouv.fr/sites/default/files/ministere-armees/Presentation%20Climate%20ans%20defence%20strategy.pdf [2023-04-21].

Frey, U. and Burgess, J. 2022. *Why Do Climate Change Negotiations Stall? Scientific Evidence and Solutions for Some Structural Problems*. Bristol University Press.

Friedman, L. 2023. Many Young Voters Bitter over Biden's Support of Willow Oil Drilling. *The New York Times*. 24 April 2023.

Foresight. 2011. *Foresight: Migration and Global Environmental Change Final Project Report*. London: The Government Office for Science.

Freestone, D. and Cicek, D. 2021. *Legal Dimensions of Sea Level Rise: Pacific Perspectives*. The World Bank & GFDRR.

Gallo, M. Sheikh, Z. and Qazi, K. 2023. Climate Crisis in Pakistan: Voices from the Ground. *Islamic Relief Canada*.

Garric, A. 2022. French Presidential Election 2022: Climate Change Is Important to Voters, but Largely Absent from Debates. *Le Monde*. 10 April 2022.

Germanwatch. 2023. *Global Climate Risk Index*. [online] Available at: https://www.germanwatch.org/en/cri [2023-10-09].

Global Methane Pledge. 2021. *Fast Action on Methane to Keep a 1.5°C Future within Reach*. [online] Available at: https://www.globalmethanepledge.org/ [2023-06-21].

Grainger, S. and Conway, D. 2014. Climate Change and International River Boundaries: Fixed Points in Shifting Sands. *WIREs Climate Change*, Vol. 5. No. 6. Pp. 835–848.

Guo, S. 2020. The Legacy Effect of Unexploded Bombs on Educational Attainment in Laos. *Journal of Development Economics*, Vol. 147.

Hadley, G. 2023. F-35s Deploy to Greenland for First Time, Operate from Thule. *Air & Space Forces Magazine*, 31 January 2023.

Hanlon, R. 2016. Freedom from Fear, *Freedom from Want: An Introduction to Human Security*. Toronto: University of Toronto Press.

Hasemyer, D. 2019. U.S. Soldiers Falling Ill, Dying in the Heat as Climate Warms. *Inside Climate News*. 23 July 2019.

Harris, P. 2016. *Global Ethics and Climate Change*. Edinburgh: Edinburgh University Press.

Heathershaw, J. 2008. Unpacking the Liberal Peace: The Dividing and Merging Peacebuilding Discourses. *Millennium: Journal of International Studies*, Vol. 36. No. 3. Pp. 597–621.

Hendrix, C. and Haggard, S. 2015. Global Food Prices, Regime Type, and Urban Unrest in the Developing World. *Journal of Peace Research*, Vol. 52. No. 2.
Henriques, M. 2020. The Rush to Claim an Undersea Mountain Range. *BBC*. 23 July 2020.
Herre, B. 2022. The World Has Recently Become Less Democratic. *Our World in Data*. 6 September 2022.
Hill, A. 2022. *COP27 Didn't Make Enough Progress to Prevent Climate Catastrophe*. Council on Foreign Relations. 21 November 2022.
Hooijer, A. and Vernimmen, R. 2021. Global LiDAR Land Elevation Data Reveal Greatest Sea-Level Rise Vulnerability in the Tropics. *Nature Communications*. Vol. 12.
HRC (UN Human Rights Committee). 2020. *Views Adopted by the Committee under Article 5 (4) of the Optional Protocol, Concerning Communication No. Communication No. 2918/2016*. CCPR/C/130/D/2918/2016. 28 December 2020.
HRW (Human Rights Watch). 2023. *Pakistan: Afghans Detained, Faced Deportation*. 31 October 2023.
Humpert, M. 2021. China to Build New Heavy Icebreaker and Lift Vessel for Arctic. *High North West*. 16 November 2021.
Humpert, M. 2023. Russia's Northern Sea Route Sees More Traffic Despite War and Sanction. *High North West*. 18 January 2023.
Iannizzi, G. 2021. *The Long-Lasting Border Issue on the Mont Blanc*. Swiss Diplomacy Student Association. [online] Available at: https://sdsa-geneve.ch/index.php/en/2021/01/16/the-long-lasting-border-issue-on-the-mont-blanc-2/ [2023-05-16].
ICJ (International Court of Justice). 1999. *Case Concerning Kasikili/Sedudu Island (Botswana/Namibia)* ICJ Reports 1999. The Hauge. 13 December 1999.
ICJ (International Court of Justice). 2005. *Case Concerning the Frontier Dispute (Benin/Niger)*. ICJ Reports 2005. The Hauge. 12 July 2005.
ICJ (International Court of Justice). 2011. *Certain Activities Carried Out by Nicaragua in the Border Area (Costa Rica V. Nicaragua)*. The Hauge.
ICJ (International Court of Justice). 2013. *Certain Activities Carried Out by Nicaragua in the Border Area (Costa Rica v. Nicaragua) Construction of a Road in Costa Rica along the San Juan River (Nicaragua v. Costa Rica)*. General list Nos. 150 and 152. The Hauge. 22 November 2023.
ICWA (Indian Council of World Affairs). 2023. *Japanese Security Policy in the East China Sea*. 31 January 2023.
IDMC (Internal Displacement Monitoring Center). 2021a. *Internal Displacement in a Changing Climate*. [online] Available at: https://www.internal-displacement.org/sites/default/files/publications/documents/grid2021_idmc.pdf
IDMC (Internal Displacement Monitoring Center). 2021b. *A Decade of Displacement in the Middle East and North Africa*. [online] Available at: https://www.internal-displacement.org/sites/default/files/publications/documents/IDMC_MenaReport_final.pdf
IDMC (Internal Displacement Monitoring Center). 2022. *Global Report on Internal Displacement 2022*. [online] Available at: https://www.internal-displacement.org/global-report/grid2022/#part1

IDMC (Internal Displacement Monitoring Center). 2023. *Global Report on Internal Displacement*. [online] Available at: https://www.internal-displacement.org/global-report/grid2023/ [2023-11-06].

IEA (International Energy Agency). 2021a. *Methane and Climate Change*. [online] Available at: https://www.iea.org/reports/methane-tracker-2021/methane-and-climate-change [2023-08-30].

IEA (International Energy Agency). 2021b. *Net Zero by 2050*. [online] Available at: https://www.iea.org/reports/net-zero-by-2050 [2023-09-04].

IEA (International Energy Association). 2021c. *Hydropower Special Market Report*. Paris: IEA.

IEA (International Energy Agency). 2022a. *CO^2 Emissions in 2022*. Paris: IEA.

IEA (International Energy Agency). 2022b. *Coal in Net Zero Transitions: Strategies for Rapid, Secure, and People-Centred Change*. World Energy Outlook Special Report.

IISS. (The International Institute for Strategic Studies). 2022. Chapter Two: Defence and Military Analysis. *Military Balance*. Vol.122. Pp. 14–25.

ILO (International Labour Organization). 1989. *C169: Indigenous and Tribal Peoples Convention, 1989* (No.169).

ILO (International Labour Organization). 2019. *Working on a Warmer Planet: The Impact of Heat Stress on Labour Productivity and Decent Work*. Geneva: ILO.

ILO (International Labour Organization). 2024. *Ratifications of C169: Indigenous and Tribal Peoples Convention, 1989*.

IMF (International Monetary Fund). 2019. *Building Resilience in Developing Countries Vulnerable to Large Natural Disasters*. IMF Policy Paper. June, 2019.

InfluenceMap. 2022. *2022 Rankings: The World's Most Obstructive Companies on Climate Policy*. Press Release. 3 November 2022.

IOM (International Organization for Migration). 2019. *Glossary on Migration*. International Migration Law No. 34. Geneva.

IOM (International Organization for Migration). 2023. *Impact of Storm Daniel*. [online] Available at: https://reliefweb.int/report/libya/libya-impact-storm-daniel-update-displacement-and-needs-november-2023 [2023-11-17].

IPCC (Intergovernmental Panel on Climate Change). 1992. *Climate Change: The 1990 and 1992 IPCC Assessments*.

IPCC (Intergovernmental Panel on Climate Change). 2018. Summary for Policymakers. In *Global Warming of 1.5°C: An IPCC Special Report on the Impacts of Global Warming of 1.5°C above Pre-industrial Levels and Related Global Greenhouse Gas Emissions Pathway, in the Context of Strengthening the Global Response to the Threat of Climate Change, Sustainable Development, and Efforts to Eradicate Poverty*. Cambridge: Cambridge University Press.

IPCC (Intergovernmental Panel on Climate Change). 2019a. Desertification. In *Climate Change and Land: An IPCC Special Report on Climate Change, Desertification, Land Degradation, Sustainable Land Management, Food Security, and Greenhouse Gas Fluxes in Terrestrial Ecosystems*. Cambridge: Cambridge University Press. Pp. 249–343.

IPCC (Intergovernmental Panel on Climate Change). 2019b. *Special Report on the Ocean and Cryosphere in a Changing Climate*. Cambridge: Cambridge University Press.

IPCC (Intergovernmental Panel on Climate Change). 2019c. Food Security. In *Climate Change and Land: An IPCC Special Report on Climate Change, Desertification, Land Degradation, Sustainable Land Management, Food Security, and Greenhouse Gas Fluxes in Terrestrial Ecosystems*. Cambridge: Cambridge University Press. Pp. 437-550.
IPCC (Intergovernmental Panel on Climate Change). 2019d. Sea Level Rise and Implications for Low-Lying Islands, Coasts and Communities. In *IPCC Special Report on the Ocean and Cryosphere in a Changing Climate*. Cambridge University Press, Cambridge and New York. Pp. 321-445.
IPCC (Intergovernmental Panel on Climate Change). 2021. Summary for Policymakers. In: *Climate Change 2021: The Physical Science Basis*. Cambridge: Cambridge University Press.
IPCC (Intergovernmental Panel on Climate Change). 2022a. Climate Change 2022: Impacts, Adaptation and Vulnerability. *Contribution of Working Group II to the Sixth Assessment Report of the Intergovernmental Panel on Climate Change*. Cambridge: Cambridge University Press. Pp. 551-712.
IPCC (Intergovernmental Panel on Climate Change). 2022b. *Fact Sheet – Africa*. Sixth Assessment Report. Working Group II-Impacts, Adaptation and Vulnerability.
IPCC (Intergovernmental Panel on Climate Change). 2023a. *Synthesis Report of the IPCC Sixth Assessment Report (AR6)*. Cambridge: Cambridge University Press.
IPCC (Intergovernmental Panel on Climate Change). 2023b. *About the IPCC*. [online] Available at: https://www.ipcc.ch/about/ [2023-11-30].
ISD (Institute for Strategic Dialogue). 2023. *Deny, Deceive, Delay. Vol. 2: Exposing New Trends in Climate Mis-and Disinformation at COP27*.
ITLOS (International Tribunal for the Law of the Sea). 2023. *Dispute Concerning Delimitation of the Maritime Boundary between Mauritius and Maldives in the Indian Ocean*. 28 General Case No. 28. 28 April 2023.
Jack, C. 2017. Lexicon of Lies: Terms for Problematic Information. *Data & Society*, Vol. 3. No. 22.
Jack, I. 2016. We Called it Racism, Now it Is Nativism. The Anti-immigrant Sentiment Is Just the Same. *The Guardian*. 12 November 2016.
Jackson, P. 2007. *From Stockholm to Kyoto: A Brief History of Climate Change*. United Nations Chronicle.
Jacobson, M. 2019. The Health and Climate Impacts of Carbon Capture and Direct Air Temperature. *Energy & Environmental Science*, No. 12.
Jarvis, T. 2014. *Contesting Hidden Waters: Conflict Resolution for Groundwater and Aquifers*. Oxon: Routledge.
Jones, N., Verkuijl, C., Muñoz Cabré, M. and Piggot, G. 2023. *Connecting the Dots: Mapping References to Fossil Fuel Production in National Plans under the UNFCCC for the 2023 Global Stocktake*. SEI Report. June 2023.
Joshi, P. 2017. The Battle for Siachen Glacier: Beyond Just a Bilateral Dispute. *Strategic Analysis*, Vol. 41. No. 5.
Jägerskog, A. and Swain, A. 2016. *Water, Migration, and How They Are Interlinked*. Working paper 27. Stockholm: SIWI.

Jägerskog, A. and Swain, A. 2024. Water, Migration, and Development. In Hellberg, S. Söderbaum, F. Swain, A. and Öjendal, J. (eds). *Routledge Handbook of Water and Development*. Oxon: Routledge.

Katz, G. and Levin, I. 2015. The Dynamics of Political Support in Emerging Democracies: Evidence from a Natural Disaster in Peru. *International Journal of Public Opinion Research*. Vol. 28. No. 2.

Kennedy, R. 2021. 'Dangerous and Delusional': Critics Denounce Saudi Climate Plan. *Al Jazeera*. 26 October 2021.

Kim, Y.-H. Min, S.-K. Gillett, N. Notz, D. and Malinina, E. 2023. Observationally-constrained Projections of an Ice-free Arctic Even under a Low Emission Scenario. *Nature Communications*. Vol. 14. No. 3139.

Klare, M. T. 2019. *All Hell Breaking Loose: The Pentagon's Perspective on Climate Change*. Audiobook.

Koks, E. van Ginkel, K. van Marle, M. and Lemnitzer, A. 2022. Brief Communication: Critical Infrastructure Impacts of the 2021 Mid-July Western European Flood Event. *Natural Hazards and Earth System Sciences*, No. 22. Pp. 3831-3838.

Krampe, F. and de Coning, C. 2021. Russia's 'Nyet' Does Not Mean Climate Security Is off the Security Council Agenda. *SIPRI*. 13 December 2021.

Kulp, S. and Strauss, B. 2019. New Elevation Data Triple Estimates of Global Vulnerability to Sea-Level Rise and Coastal Flooding. *Nature Communications*. Vol. 20. No. 4844.

Kumar, B, 2023. Accountable Allies: The Undue Influence of Fossil Fuel Money in Academia. *Data for Progress*. Available at: https://www.filesforprogress.org/memos/accountable-allies-fossil-fuels.pdf [2023-08-31].

Kuper, S. 2023. The Red-Hot Issue in the Spanish Elections Should Be Climate, Not Culture Wars. *Financial Times*. 27 July 2023.

Kurbanov, A. and Prokhoda, V. 2019. Ecological Culture: an Empirical Projection: Attitudes of Russians towards Climate Change. *Monit Public Opin Econ Soc Chang*, Vol. 4. Pp. 347-370.

L.A. Times Archives. 1998. Argentina, Chile Sign Pact to End Their Border Dispute. *Los Angeles Times*. 17 December 1998.

Lee, J. and Tanaka, K. 2016. Climate Change, Conflict, and Moving Borders. *The International Journal of Climate Change: Impacts and Responses*. Vol. 8. Pp. 29-44.

Legal Information Institute. 2021. *American Clean Energy and Security Act of 2009*. Cornell Law School. [online] Available at: https://www.law.cornell.edu/wex/american_clean_energy_and_security_act_of_2009 [2023-09-11].

Li, J., Yang, J., Liu, M., Ma, Z., Fang, W. and Bi, J. 2022. Quality Matters: Pollution Exacerbates Water Scarcity and Sectoral Output Risks in China. *Water Research*, Vol. 224.

Lin, T-H. 2015. Governing Natural Disasters: State Capacity, Democracy, and Human Vulnerability. *Social Forces*, Vol. 93. No. 3.

Lindvall, D. 2021. *Democracy and the Challenge of Climate Change*. International IDEA Discussion Paper 3/2021.

Liptak, K. and Nilsen, E. 2022. Biden Says US Is Back as a Leader on Fighting Climate Change as He Urges All Nations to Step up Their Ambitions. *CNN*. 11 November 2022.

LKAB. 2023. *Europe's Largest Deposit of Rare Earth Metals Is Located in the Kiruna Area.* 12 January 2023.
LRF (Lloyd's Register Foundation). 2022. *World Risk Poll 2021: A Changed World?* [online] Available at: https://wrp.lrfoundation.org.uk/LRF_2021_report_risk-in-the-covid-age_online_version.pdf [2023-08-21].
Maldives Financial Review. 2023. *Boundary Dispute between Mauritius and Maldives.* 29 April 2023.
Manzano Iturra, K. 2019. *Campos de Hielo Sur. Controversias en Torno a la Frontera Chileno-Argentina (1990–2012).* Politica y Estrategia. No. 134. ISSN 0719-8027.
Marshak, S. 2019. *Earth: Portrait of a Planet.* 6th edition. New York: W.W. Norton & Company.
Maurer, J. Schaefer, J. Rupper, S. and Corley, A. 2019. Acceleration of Ice Loss across the Himalayas over the Past 40 Years. *Science Advances.* Vol. 5. No. 6.
McCarthy, N. 2019. Oil and Gas Giants Spend Millions Lobbying to Block Climate Change Policies. *Forbes.* 25 March 2019.
McCormick, E. 2022. Patagonia's Billionaire Owner Gives Away Company to Fight Climate Crisis. *The Guardian.* 15 September 2022.
McKay, D. Staal, A. Abrams, J. Winkelmann, R. Sakschewski, B. Loriani, S. Fetzer, I. Cornell, E. Rockström, J. and Lenton, M. 2022. Exceeding 1.5°C Global Warming Could Trigger Multiple Climate Tipping Points. *Science.* No. 377.
McKie, R. 2019. Climategate 10 Years on: what Lessons Have We Learned? *The Guardian.* 9 November 2019.
McSweeney, R. 2023. Analysis: Which Countries Have Sent the Most Delegates to COP28? *CarbonBrief.*
Meadows, D., Meadows, D., Randers, J. and Behrens, W. 1972. *The Limits to Growth: A Report Got the Club of Rome's Project on the Predicament of Mankind.* Potomac.
Meijers, M., Drunen, Y. and Jacobs, K. 2022. It's a Hoax! the Mediating Factors of Populist Climate Policy Opposition. *West European Politics,* Vol. 26. No. 7.
Meng, K. and Rode, A. 2019. The Social Cost of Lobbying over Climate Policy. *Nature Climate Change.* Vol. 9. Pp. 472-476.
Michaelowa, A., Koch, T., Charro, D. and Gameros, C. 2022. *Military and Conflict-Related Emissions: Kyoto to Glasgow and beyond.* Freiburg: Perspectives Climate Research.
Miljand, M. 2022. (cited by Oidermaa, M.) *Konkreta Klimatförslag Saknades I Valdebatten.* Stockholm University. 16 September 2022.
Ministerio de Defensa Nacional. 2017. *Libro de la Defensa Nacional de Chile.* [online] Available at: https://www.defensa.cl/media/LibroDefensa.pdf [2023-04-28].
Ministry of Foreign Affairs (Kingdom of Thailand). 2021. *Deputy Prime Minister and Minister of Foreign Affairs and the Minister of Foreign Affairs of the Lao PDR Discussed Ways to Strengthen Cooperation in the Border Area.* [online] Available at: https://www.mfa.go.th/en/content/thailaojbcdiscussion05112564-2?page=5d5bd3cb15e39c306002a9ac&menu=5d5bd3dc15e39c306002ab1c [2023-05-09].
Mitchell, S. and Zawahri, N. 2015. The Effectiveness of Treaty Design in Addressing Water Disputes. *Journal of Peace Research,* Vol. 52. No. 2.

Methmann, C. and Oels, A. 2015. From 'fearing' to 'empowering' Climate Refugees: Governing Climate-Induced Migration in the Name of Resilience. *Security Dialogue*, Vol. 46. No. 1.

Moens, J. 2022. Hurricane Michael Hit Florida Panhandle in 2018 with 155 MPH Winds. Some Black Low-Income Neighborhoods Still Haven't Recovered. *Inside Climate News*. 22 March 2022.

Mohamed, M. 2020. An Assessment of Forest Cover Change and its Driving Forces in the Syrian Coastal Region during a Period of Conflict, 2010 to 2020. *Land*, Vol. 2. No. 191.

Moore, S. and Melton, M. 2019. China's Pivot on Climate Change and National Security. *Lawfare*. 2 April 2019.

Moritaka, H. 2013. Islands' Sea Areas: Effects of a Rising Sea Level. *Review of Island Studies*. Translated from 'shima no kaiiki to kaimen ojsho'. *Tossho Kenkyu Journal, OPRF Center for Island Studies*, Vol. 2. No. 1. Pp. 74–87.

Mounk, Y. 2018. *The People vs. Democracy: Why Our Freedom Is in Danger and How to Save It*. Cambridge: Harvard University Press.

Mudde, C. 2012. *The Relationship between Immigration and Nativism in Europe and North America*, Washington DC: Migration Policy Institute.

Mueller, V., Gray, C. and Kosec, K. 2014. Heat Stress Increases Long-Term Human Migration in Rural Pakistan. *Nature Climate Change*. Vol. 4. Pp. 182–185.

Müller, J. 2016. *What Is Populism?* Philadelphia: University of Pennsylvania Press.

Mujuthaba, A. and Brewster, D. 2021. *Maldives Embroiled in Mauritius-UK Tussle over Chagos*. Australian National University.

Myeni, T. 2022. What Is Operation Dudula, South Africa's Anti-migration Vigilante. *Al-Jazeera*. 8 April 2022.

Nabi, I. 2023. *Responding to Pakistan Floods*. Brookings. 10 February 2023. [online] Available at: https://www.brookings.edu/articles/pakistan-floods/ [2023-09-29].

NASA. 2019a. *A Degree of Concern: Why Global Temperatures Matter*. [online] Available at: https://climate.nasa.gov/news/2865/a-degree-of-concern-why-global-temperatures-matter/ [2023-06-27].

NASA. 2019b. *The Atmosphere: Getting a Handle on Carbon Dioxide: Sizing up Humanity's Impacts on Earth's Changing Atmosphere: A Five-Parts Series*. [online] Available at: https://climate.nasa.gov/news/2915/the-atmosphere-getting-a-handle-on-carbon-dioxide/ [2023-04-14].

NASA. 2023a. *Tracking Canada's Extreme 2023 Fire Season*. [online] Available at: https://earthobservatory.nasa.gov/images/151985/tracking-canadas-extreme-2023-fire-season [2023-11-28].

NASA. 2023b. *Ocean Warming*. [online]. Available at: https://climate.nasa.gov/vital-signs/ocean-warming/ [2023-05-31].

NASA. nd. *Understanding Sea Level*. [online] Available at: https://sealevel.nasa.gov/understanding-sea-level/key-indicators/steric-height [2023-05-31].

National Geographic Society. 2022a. *Towers*. [online] Available at: https://education.nationalgeographic.org/resource/water-towers/ [2023-05-17].

National Geographic Society. 2022b. *Arctic*. [online] Available at: https://education.nationalgeographic.org/resource/arctic/ [2023-05-23].

National Intelligence Council. 2021. *Climate Change and International Responses Increasing Challenges to US National Security through 2040*. National Intelligence Estimate NIC-NIE-2021-10030-A.

National Park Service. nd. *The Flowing Border*. [online] Available at: https://www.nps.gov/cham/learn/nature/upload/River-Movements-copier-no-bleed.pdf [2023-05-04].

NATO. 2022a. *Opening Speech by NATO Secretary General Jens Stoltenberg at the High-Level Dialogue on Climate and Security*, NATO Public Forum. June 28 2022. [online] Available at: https://www.nato.int/cps/en/natohq/197168.htm?selectedLocale=

NATO. 2022b. *Climate Change & Security Impact Assessment*. The Secretary General's Report. [online] Available at: https://www.nato.int/nato_static_fl2014/assets/pdf/2022/6/pdf/280622-climate-impact-assessment.pdf [2023-04-18].

NATO. 2023. *Joint Force Development Experimentation & Wargaming Branch 2023 Fact Sheet – Nato Operational Energy Concept*. NATO Allied Command Transformation.

Naturvårdsverket. 2021. *Allmänhetens Kunskap Och Attityder till Klimatfrågor*. [online] Available at: https://www.naturvardsverket.se/amnesomraden/klimatomstallningen/sveriges-klimatarbete/allmanhetens-kunskap-och-attityder-till-klimatfragor/[2023-09-08].

NCEI (National Centers for Environmental Information). 2023. *Assessing the Global Climate in 2022*. [online] Available at: https://www.ncei.noaa.gov/news/global-climate-202212 [2023-11-28].

ND-GAIN (Notre Dame Global Adaptation Initiative). 2023. *Rankings*. [online] Available at: https://gain.nd.edu/our-work/country-index/rankings/ [2023-09-21].

Nilsen, E. 2023. The Willow Project Has Been Approved. Here's what to Know about the Controversial Oil-Drilling Venture. *CNN*. 13 March 2023.

NOAA (National Oceanic & Atmospheric Administration). 2016. *JPSS Satellites Help Ships Navigate the Northwest Passage*. 4 October 2016.

NOAA (National Oceanic and Atmospheric Administration). 2019. *Coral Reef Ecosystems*. [online] Available at: https://www.noaa.gov/education/resource-collections/marine-life/coral-reef-ecosystems [2023-11-03].

Notz, D. and SIMIP Community. 2020 Arctic Sea Ice in CMIP6. *Geophysical Research Letters*, No. 47.

OECD (Organization for Economic Co-operation and Development). 2022. *Climate Finance provided and Mobilised by Developed Countries in 2016–2020*.

OHCHR (Office of the United Nations High Commissioner for Human Rights). 2023. *Ref.: AL OTH 53/2023*. Communication to Mr Nasser (Preseident and CEO of Aramco). 26 June 2023.

Ogata, S. N. and Sen, A. 2003. *Final Report of the Commission on Human Security*. New York: CHS Secretariat.

Onti, T. and Schulte, L. 2012. Soil Carbon Storage. *Nature Education Knowledge*, Vol. 3. No. 10.

Oregon State University. (2024). *Transboundary Freshwater Diplomacy Database (TFDD)*. Additional information on the TFDD can be found at: http://transboundarywaters.science.oregonstate.edu

Orlove, B., Wiegandt, E. and Luckman, B. H. 2008. *Darkening Peaks: Glacier Retreat, Science and Society*. Berkeley: University of California Press.

Osaka, S. 2023. The Willow Oil Project Debate Comes Down to This Key Climate Change Question. *Washington Post.* 17 March 2023.
O'Toole, G. 2017. *Environmental Security in Latin America.* Oxfordshire: Routledge.
Ottaway, R. 2021. *Saudi Arabia's Green Initiative Aims to Exonerate Fossil Fuel Advocacy.* Wilson Center. 1 November 2021.
Our World in Data. 2021. *Share of Global CO_2 Emissions and Population.*
Ovaska, M. Nasralla, S. and Abnett, K. 2021. *Who Is the Biggest Polluter? Depends How You Ask.* Reuters Graphics.
Oxfam. 2023. *Oxfam Reaction to UN Report on Adaptation Finance.* 2 November 2023.
Oxfam. 2024. *Inequality Inc: How Corporate Power Divides Our World and the Need for a New Era of Public Action.* Oxford: Oxfam International.
Oxfam and SEI (Stockholm Environment Institute). 2023. *Climate Equality: A Planet for the 99 %.* Oxford: Oxfam International.
Pacific Island Forum. 2021. *Declaration on Preserving Maritime Zones in the Face of Climate Change-Related Sea-Level Rise.* 6 August 2021.
Paddison, L. 2024. Norway Parliament Approves Highly Controversial Deep Sea Mining. *CNN.* 9 January 2024.
Panday, A. 2021. *Melting Glaciers, Threatened Livelihoods: Confronting Climate Change to Save the Third Pole.* Policy Brief, UNDP Regional Bureau for Asia and the Pacific Strategy, Policy and Partnership.
Parkinson, S. 2020. *The Carbon Boot-Print of the Military. Responsible Science.* No.2. SGR: Scientists for Global Responsible Science.
Parkinson, S. and Cottrell, L. 2021. *Under the Radar: The Carbon Footprint of Europe's Military Sectors: A Scoping Study.* SGR: Scientists for Global Responsibility and COEBS: Conflict and Environment Observatory.
Parkinson, S. and Cottrell, L. 2022. *Estimating the Military's Global Greenhouse Gas Emissions.* SGR: Scientists for Global Responsibility and COEBS: Conflict and Environment Observatory.
Parry, J.-E. and Terton, A. 2016. *How Are Vulnerable Countries Adapting to Climate Change?* International Institute for Sustainable Development. 21 November 2016.
Patagonia. 2023. *Earth Is Now Our Only Shareholder.* [online] Available at: https://www.patagonia.com/ownership/ [2023-12-01].
Pelet, V. 2016. Puerto Rico's Invisible Health Crisis. *The Atlantic.* September 3, 2016.
Pereira, P. Bašić, F. Bogunovic, I. and Barcelo, D. 2022. Russian-Ukrainian War Impacts the Total Environment. *Science of the Total Environment,* No. 837.
PIB Delhi. 2022. *Melting of Himalayan Glaciers.* 6 April 2022.
Pinson, A. White, S. Moore, D. Samuelson, B. Thames, P. O'Brien, C. Hiemstra, P. Loechl, E. and Pitchie, E. 2020. *Army Climate Resilience Handbook.* Washington, DC: U.S. Army Crops of Engineers.
Phillips, A. 2022a. Biden Pulls 3 Offshore Oil Lease Sales, Curbing New Drilling This Year. *Washington Post.* 12 May 2022.
Phillips, L. 2022b. Why Rising Waters Doesn't Need to Mean Retreating Borders: Protecting the Economic Rights of Island Nations. *Asia & the Pacific Policy Society: Development, Environment & Energy, Law.* 12 August 2022.

Pokharel, T., Manandhar, M., Dahal, A., Chalise, B., Bhandari, R. and Kharel, T. 2018. Political Economy Analysis of Post-Earthquake Reconstruction in Nepal: An Assessment of Emerging Role of Local Governments. *Nepal Administrative Staff College and the Asia Foundation*. September 2018.

Pompeo, M. 2019. *Looking North: Sharpening America's Arctic Focus*. 6 May 2019.

Popelka, S. and Smith, L. 2020. Rivers as Political Borders: a New Subnational Geospatial Dataset. *Water Policy*, Vol. 22. No. 3.

Porterfield, C. 2022. Patagonia Founder Gives Away Entire Company to Fight Climate Change. *Forbes*. 14 September 2022.

Prause, L. 2020. Chapter 10: Conflicts Related to Resources: The Case of Cobalt Mining in the Democratic Republic of Congo. In Bleicher, A. and Pehlken, A. (eds), *The Material Basis of Energy Transition*. London: Academic Press.

PRC (Pew Research Center). 2016. *Number of Refugees to Europe Surges to Record 1.3 Million in 2015*. 2 August 2016.

Proposition. 1983/84:202. *Om ändring av riksgränsen mellan Sverige och Finland*.

PwC. 2022. *Klimatpolitiken viktig när svenskarna lägger sina röster- men få är redo att minska resorna för att skona klimatet*. 9 September 2022.

Quell, M. 2023. *UN Tribunal Divides Contested Ocean Territory between Mauritius and Maldives*. Courthouse News Service. 28 April 2023.

Rahman, M., Anbarci, N., Bhattacharya, P. and Ulubasoğlu, M. 2017. Can Extreme Rainfall Trigger Democratic Change? the Role of Flood-Induced Corruption. *Public Choice*. No. 171.

Rahman, M., Anbarci, N. and Ulubasoğlu, M. 2022. 'Storm Autocracies': Islands as Natural Experiments. *Journal of Development Economics*, Vol. 159.

Rajaeifar, M., Belcher, O., Parkinson, S., Neimark, B., Weir, D., Ashworth, K., Larbi, R. and Heidrich, O. 2022. Decarbonize the Military – Mandate Emissions Reporting. *Nature*, Vol. 611. Pp. 29-32.

Raleigh, C., Jordan, L. and Salehyan, I. 2008. *Assessing the Impact of Climate Change on Migration and Conflict*. Washington, DC: The World Bank.

Rantanen, M., Karpechko, A., Lipponen, A., Nordling, K., Hyvärinen, O., Ruosteenoja, K., Vihma, T. and Laaksonen, A. 2022. The Arctic Has Warmed Nearly Four Times Faster Than the Globe since 1979. *Communications Earth & Environment*, Vol. 9. No. 168.

Rasheed, A. 2022. Greening National Security Policy in the Indo-Pacific. *Asia & the Pacific Policy Society*. August 8 2022.

Rawtani, D., Gupta, G., Khatri, N., Rao, P. and Hussain, C. 2022. Environmental Damages Due to War in Ukraine: A Perspective. *Science of the Total Environment*, No. 850.

Regeringskansliet. 2006. *Ny gränsöverenskommelse med Finland*. [online] Available at: https://www.regeringen.se/rattsliga-dokument/lagradsremiss/2006/02/ny-gransalvsoverenskommelse-med-finland/ [2023-05-04].

Reuters. 2021. *Developer Officially Cancels Keystone XL Pipeline Project Blocked by Biden*. 10 June 2021.

Reuters. 2023. *Factbox: What Is the Willow Project in Alaska, and Why Do Green Activists Oppose it?* 15 March 2023.

Rhode, R. 2024. *Global Temperature Report for 2023*. Berkeley Earth. 12 January 2024.

Rick, B., McGrath, D., McCoy, S. and Armstrong, W. 2023. Unchanged Frequency and Decreasing Magnitude of Outbursts from Ice-Dammed Lakes in Alaska. *Nature Communications*, Vol. 14.

Rising. M. 2023. *Norsk ovädersnota bland de dyraste någonsin.* Dagens Industri. 16 October 2023.

Robinson, M. 2021. Hope at COP26 Must Be Backed by Decisive Action from World Leaders. *The Elders.* [online] Available at: https://theelders.org/news/hope-cop26-must-be-backed-decisive-action-world-leaders [2024-03-11].

Rosignoli, F. 2022. *Environmental Justice for Climate Refugees.* Oxon: Routledge.

Ryan, M. 2023. U.S. Seeks to Expand Developing World's Influence at United Nations. *Washington Post.* 12 June 2023.

Räsänen, T., Varis, O., Scherer, L. and Kummu, M. 2018. Greenhouse Gas Emissions of Hydropower in the Mekong River Basin. *Environmental Research Letters,* Vol. 13.

Salido, A. 2024. *Gardi Sugdub: The Americas' Disappearing Island.* BBC. 5 January 2024.

Sanchez, F. Nykvist, P. Olsson, O. and Linde, L. 2023. *Lessons from Oil and Gas Transition in the North Sea.* SEI Report. Stockholm: Stockholm Environment Institute.

Schaeffer, K. and van Green, T. 2022. *Key Facts about U.S. Voter Priorities Ahead of the 2022 Midterm Elections.* Pew Research Center. 3 November 2022.

Schoolmeester, T., Johansen, K., Alfthan, B., Baker, E., Hesping, M. and Verbist, K. 2018. *The Andean Glacier and Water Atlas: The Impact of Glacier Retreat on Water Resources.* UNESCO and GRID-Arendal.

Schwartz, P. and Randall, D. 2003. An Abrupt Climate Change Scenario and Its Implications for United States National Security. *Pentagon Study,* October 2003.

SEI. IISD. ODI. E3G. and UNEP. 2021. *The Production Gap Report 2021: Governments' Planned Fossil Fuel Production Remains Dangerously Out of Sync with Paris Agreement Limits.* [online] Available at: https://productiongap.org/wp-content/uploads/2021/11/PGR2021_web_rev.pdf [2023-08-09].

Shkaki, J. 2020. Turkish-Backed Militias Cut Down Half a Million Olive Trees in Syria's Afrin-Local Monitor. North Press Agency. 19 August 2020.

SIPRI (Stockholm International Peace Research Institute). 2023. *World's Military Expenditure Reaches New Record High as European Spending Surges.*

Smith, C., Baker, J. and Spracklen, D. 2023. Tropical Deforestation Causes Large Reductions in Observed Precipitation. *Nature.* Vol. 615. Pp. 270-275.

Sowers, J., Weinthal, E. and Zawahri, N. 2017. Targeting Environmental Infrastructures, International Law, and Civilians in the New Middle Eastern Wars. *Security Dialogue,* Vol. 48. No. 5. Pp. 410-430.

Sphere India. 2023. *Multi-Stakeholder Consultation: Flash Floods in Sikkim and West Bengal.* Summary Report. 20 October 2023.

Stokes, B. 2018. *Populist Views in Europe: It's Not Just the Economy.* Pew Research Center. 19 July 2018.

Supran, G. and Oreskes, N. 2017. Assessing ExxonMobil's Climate Change Communications (1977-2014). *Environmental Research Letters,* Vol. 12. No. 8.

Supran, G., Rahmstorf, S. and Oreskes, N. 2023. Assessing ExxonMobil's Global Warming Projections. *Science.* Vol. 379. No. 6628.

Swain, A. 1996a. Environmental Migration and Conflict Dynamics: Focusing on Developing Regions. *Third World Quarterly*. Vol. 17. No. 5. Pp. 959-973.
Swain, A. 1996b. *The Environmental Trap: The Ganges River Diversion, Bangladeshi Migration and Conflicts in India*. Report No. 41. Department of Peace and Conflict Research. Sweden: Uppsala University.
Swain, A. 2009. The Indus II and Siachen Peace Park: Pushing the India-Pakistan Peace Process Forward. *The Commonwealth Journal of International Affairs*. Vol. 98. No. 404.
Swain, A. 2010. *Struggle against the State: Social Network and Protest Mobilization in India*. London: Routledge.
Swain, A., Bali Swain, R., Themnér, A. and Krampe, F. 2011. *Climate Change and the Risk of Violent Conflicts in Southern Africa*. Pretoria: Global Crisis Solutions.
Swain, A. 2012a. Global Climate Change and Challenges for International River Agreements. *International Journal of Sustainable Society*, Vol. 4. Pp. 72-87.
Swain, A. 2013. *Understanding Emerging Security Challenges: Threats and Opportunities*. London: Routledge.
Swain, A. 2019. *Increasing Migration Pressure and Rising Nationalism: Implications for Multilateralism and SDG Implementation*, Background Paper Prepared for the Development Policy Analysis Division of the United Nations, Department of Economics and Social Affairs. June.
Swain, A. 2015. Climate Change: Threat to National Security. In Domonic, A. Bearfield, E. Berman, M. and Dubnick, J. (eds). *Encyclopedia of Public Administration and Public Policy*. 3rd edition. London: CRC Press.
Swain, A. and Jägerskog, A. 2016. *Emerging Security Threats in the Middle East*. Lanham: Rowman & Littlefield.
Swain, A. and Öjendal, J. (eds) 2018. *Routledge Handbook of Environmental Conflict and Peacebuilding*. Oxon: Routledge.
Swain, A. 2020. Climate Change, Collective Action, and Peaceful Change. In Paul, T.V., Larson, D.W., Trinkunas, H., Wivel, A. and Emmers, R. (eds), *The Oxford Handbook of Peaceful Change in International Relations*. Oxford: Oxford University Press.
Sullivan, R., Black, R. and Kyriacou, G. 2023. What Is Climate Change Lobbying? In *The London School of Economics and Political Science and Grantham Research Institute on Climate Change and the Environment*. 17 February 2023.
SVT (Sveriges Television). 2021. *Oljebråk kan bli avgörande i norska valet*. 13 September 2021.
Tabuchi, H. 2022. Inside the Saudi Strategy to Keep the World Hooked on Oil. *The New York Times*. 21 November 2022.
Taylor, C., Robinson, T., Dunning, S., Carr, R. and Westoby, M. 2023. Glacial Lake Outburst Floods Threaten Millions Globally. *Nature Communications*, Vol. 14. No. 487.
Thangaraj, A. and Chowdhury, A. 2022. Energy, Geopolitics and the Dying Arctic Ice Fields: an Enviro-Political Perspective. In *IOP Conference Series: Earth and Environmental Science*. No. 1084.
The Economic Times. 2019. *Armored, Specialized Vehicles of Armed Forces Exempted from BS-VI Emission Norms*. 2 August 2019.

The Guardian. 2022. *Biden Accuses Oil Companies of 'war Profiteering' and Threatens Windfall Tax.* 1 November 2022.
The IMBIE Team. 2018. Mass Balance of the Antarctic Ice Sheet from 1992 to 2017. *Nature.* Vol. 558. Pp. 219–222.
The IMBIE Team. 2019. Mass Balance of the Greenland Ice Sheet from 1992 to 2018. *Nature.* Vol. 579. Pp. 233–239.
The National Autonomous University of Mexico. 2009. *Concuye La Unam Que No Exsiste La Isla Bermeja en la Ubicación Señalada.* 23 June 2009.
The White House. 2022. *National Strategy for the Arctic Region.* October 2022. [online] Available at: https://www.whitehouse.gov/wp-content/uploads/2022/10/National-Strategy-for-the-Arctic-Region.pdf [2023-05-31].
Toreti, A., Bavera, D., Acosta Navarro, J., Cammallerri, C., de Jäger, A., Di Ciollo, C., Hrast Essenfelder, A., Maetens, W., Magni, D., Masante, D., Mazzeschi, M., Niemeyer, S. and Spinoni, J. 2022. *Drought in Europe August 2022.* Luxembourg: Publications Office of the European Union.
Trump White House. 2017. *Statement by President Trump on the Paris Climate Accord.* 1 June 2017. [online] Available at: https://trumpwhitehouse.archives.gov/briefings-statements/statement-president-trump-paris-climate-accord/ [2023-08-17].
Turchetti, S. 2018. *Greening the Alliance: The Diplomacy of NATO's Science and Environmental Initiatives.* Chicago: The University of Chicago Press.
Tømmerbakke, S. 2019. *30-Year-Old Compromise Divides USA and Canada.* High North West. 12 April 2019.
UCDP (Uppsala Conflict Data Program). 2023a. *Major Data Releases and Conflict in Sudan.* Newsletter #15. UCDP.
UCDP (Uppsala Conflict Data Program). 2023b. *Number of Deaths in Armed Conflicts Have Doubled.* 13 June 2023.
Uddin, M. K. 2017. Climate Change and Global Environmental Politics: North-South Divide. *Environmental Policy and Law,* Vol. 47. Nos 3-4.
Ullman, R. 1983. Redefining Security. *International Security,* Vol. 8. No. 1. Pp. 129–153.
UK Parliament. 2024. *Chagos Islands: Security, Resettlement and Sovereignty to Be Scrutinised by FAC.* 23 February 2024.
UN. 1973. *Report of the United Nations Conference on the Human Environment.* Stockholm. 5-16 June 1972.
UN. 2007. *Security Council Holds First-Ever Debate on Impact of Climate Change on Peace, Security, Hearing over 50 Speakers.* 17 April 2007.
UN. 2011. *Security Council, in Statements, Says 'Contextual Information' on Possible Security Implications of Climate Change Important when Climate Impacts Drive Conflict.* 20 July 2011.
UN. 2019. *World Urbanization Prospect: The 2018 Revision (ST/ESA/SER.A/420).* Department of Economic and Social Affairs. New York: United Nations.
UN. 2022. *Draft Principles on Protection of the Environment in Relation to Armed Conflicts.* [online] Available at: https://legal.un.org/ilc/texts/instruments/english/draft_articles/8_7_2022.pdf [2023-04-27].
UN. 2023a. *Paris Agreement.* United Nations Treaty Collection. Status as at: 2023-12-13.

UN. 2023b. *Greenwashing-the Deceptive Tactics behind Environmental Claims.* [online] Available at: https://www.un.org/en/climatechange/science/climate-issues/greenwashing [2023-11-01].

UN. 2023c. *Glaciers Largest Freshwater Reservoir on Planet, but Threatened by Global Warming, Secretary-General Warns Event, Stressing Consequences Could Be Catastrophic.* [online] Available at: https://press.un.org/en/2023/sgsm21738.doc.htm [2023-05-10].

UN. 2023d. *The Sustainable Development Goals Report: Special Edition.* [online] Available at: https://unstats.un.org/sdgs/report/2023/The-Sustainable-Development-Goals-Report-2023.pdf [2023-10-02].

UN General Assembly. 1951. *Convention Relating to the Status of Refugees.* 28 July 1951. United Nations, Treaty Series. vol. 189, p. 137.

UN General Assembly. 1987. *Forty-second Session General Assembly.* A/42/PV.41. New York. 19 October 1987.

UN-Habitat. 2023. *Climate Change.* [online] Available at: https://unhabitat.org/topic/climate-change [2023-11-07].

UN Secretary-General. 2023. *Secretary-General's Remarks to the Security Council Debate on 'Sea-Level Rise: Implications for International Peace and Security'.* New York. 14 February 2023.

UN Women. 2022. *Explainer: How Gender Inequality and Climate Change Are Interconnected.* 28 February 2022.

UNCLOS. 1982. *United Nations Convention on the Law of the Sea.*

UN-DESA (United Nations Department of Economic and Social Affairs). 2020. *World Social Report 2020: Inequality in a Rapidly Changing World.* United Nations Publications.

UNDP (United Nations Development Program). 1994. *Human Development Report 1994.* Oxford: Oxford University Press.

UNDP (United Nations Development Program). 2017. *Zimbabwe Human Development Report.* [online] Available at: https://www.undp.org/sites/g/files/zskgke326/files/migration/zw/UNDP_ZW_2017ZHDR_Briefs–Climate-Change-and-Education.pdf [2023-10-02].

UNDP (United Nations Development Program). 2022. *Health Impacts and Social Costs Associated with Air Pollution in Larger Urban Areas in Ukraine.* [online] Available at: https://www.undp.org/ukraine/publications/health-impacts-and-social-costs-associated-air-pollution-larger-urban-areas-ukraine [2023-04-05].

UNEP (United Nations Environment Program). 2023a. *Broken Record: Temperatures Hit New Highs, yet World Fails to Cut Emissions (Again).* Nairobi: UNEP.

UNEP (United Nations Environment Program). 2023b. *Adaptation Gap Report: Underfinanced. Underprepared.* Nairobi: UNEP.

UNESCO. 2018. *The United Nations World Water Development Report 2018: Nature-Based Solutions for Water.* New York: World Water Assessment Program.

UNESCO. 2021. *Progress on Transboundary Water Cooperation.* Tracking SDG 6 Series: Global Status of SDG Indicator 6.5.2 and Acceleration Needs. New York: United Nations Publications.

UNESCO. 2022. *The United Nations World Water Development Report 2022: Groundwater: Making the Invisible Visible*. New York: World Water Assessment Program.

UNFCCC (United Nations Framework Convention on Climate Change). 1997. *Kyoto Protocol to the United Nations Framework Convention on Climate Change*. FCCC/CP/1997/L.7/Add.1. 10 December 1997.

UNFCCC (United Nations Framework Convention on Climate Change). 2015. *Paris Agreement*. [online] Available at: https://unfccc.int/sites/default/files/english_paris_agreement.pdf [2023-08-10].

UNFCCC (United Nations Framework Convention on Climate Change). 2023a. *Long-term Strategies Portal*. [online] Available at: https://unfccc.int/process/the-paris-agreement/long-term-strategies [2023-12-13].

UNFCCC (United Nations Framework Convention on Climate Change). 2023b. *Conference of the Parties Serving as the Meeting of the Parties to the Paris Agreement*. [online] Available at: https://unfccc.int/sites/default/files/resource/cma2023_L17_adv.pdf [2024-01-05].

UNFCCC (United Nations Framework Convention on Climate Change). 2023c. *COP28 Agreement Signals 'Beginning of the End' of the Fossil Fuel Era*. Press Release. 13 December 2023.

UNFCCC (United Nations Framework Convention on Climate Change). n.d. *Impacts on Climate Refugees of the Carterets Islands*, Papua New Guinea. [online] Available at: https://seors.unfccc.int/applications/seors/attachments/get_attachment?code=E08A4IBVOE450Z2EQXIPM1G3R3AO0OBM [2023-11-03].

Union of Concerned Scientists. 2007. *ExxonMobil's Disinformation Campaign*. Smoke, Mirrors and Hot Air.

Union of Concerned Scientists. 2016. *The US Military on the Front Lines of Rising Seas*.

V-Dem Institute. 2023. *Democracy Report 2023: Defiance in the Face of Autocratization*.

Van den Plas, S. and Martellucci, E. 2021. *Survival Guide to EU Carbon Market Lobby*. Carbon Market Watch. June 2021.

Veh, G., Korup, O. and Walz, A. 2019. Hazard from Himalayan Glacier Lake Outburst Floods. *Proceedings of the National Academy of Sciences*. Vol. 117. No. 2.

Vigil, S., Torre, A. and Kim, D. 2022. Exploring the Environment-Conflict-Migration Nexus in Asia. DRC Asia Climate Framework- Research Report. March 2022.

Vigna, L. and Friedrich, J. 2023. *9 Charts Explain Per Capita Greenhouse Gas Emissions by Country*. World Resources Institute. 8 May 2023.

Vilca, O., Mergili, M., Emmer, A., Frey, H. and Huggel, C. 2021. The 2020 Glacial Lake Outburst Flood Process Chain at Lake Salkantaycocha (Cordillera Vilcabamba, Peru). *Landslides*, Vol. 18. Pp. 2211-2223.

Vivar, N. 2020. *Chile y Argentina: Un conflict histórico en Campos de Hielo Sur*. Fundación Glaciares Chilenos. 21 October 2020.

Vrba, M. 2023. Climate Scepticism the Russian Way. *Green European Journal*. 13 June 2023.

Wada, Y., Lo, M., Yeh, P., Reager, J., Famiglietti, J., Wu, R. and Tseng, Y.-H. 2016a. Fate of Water Pumped from Underground and Contributions to Sea-Level Rise. *Nature Climate Change*, Vol. 6. Pp. 777-780.

Wada, Y., Flörke, M., Hanaski, N., Eisner, S., Fischer, G., Tramberend, S., Satoh, Y., van Vliet, M., Yillia, P., Ringler, C., Burek, P. and Wiberg, D. 2016b. Modelling Global Water Use for the 21st Century: The Water Futures and Solutions (WFaS) Initiative and Its Approaches. *Geoscientific Model Development*, Vol. 9. Pp. 175-222.

Wang, S., Che, Y. and Xinggang, M. 2020. Integrated Risk Assessment of Glacier Lake Outburst Flood (GLOF) Disaster over the Qinghai-Tibetan Plateau (QTP). *Landslides*, Vol. 17. Pp 2849-2863.

Walicki, N., Ioannides, M. and Tilt, B. 2017. *Dams and Internal Displacement: An Introduction*. Case Study Series, Dam Displacements. IDMC and Oregon State University.

Wallensteen, P. 2011. The Origins of Contemporary Peace Resaerch. In: Höglund, K. and Öberg, M. (eds), *Understanding Peace Research: Methods and Challenges*. London: Routledge. Pp. 14-32.

Ward, C. and Ruckstuhl, S. 2017. *Water Scarcity, Climate Change and Conflict in the Middle East: Securing Livelihoods, Building Peace*. London: I.B. Tauris. Bloomsbury Publishing Plc.

WCPFC (Western and Central Pacific Fisheries Commission). 2022. *The Western and Central Pacific Tuna Fishery: 2021 Overview and Status of Stocks*. New Caledonia: Pacific Community. 21 November 2022.

WEF (World Economic Forum). 2023a. *Closing the Gap: Accelerating Decarbonization and the Energy Transition in MENA*. Insight Report. October 2023.

WEF (World Economic Forum). 2023b. *The Global Risks Report 2023*. 18th Edition. Insight Report. Cologny/Geneva: World Economic Forum.

Weimann, G. 2004. *Cyberterrorism: How Real Is the Threat?* Washington DC: United States Institute of Peace.

Weinthal, E. and Sowers, J. 2023. Targeting Libya's Rentier Economy: The Politics of Energy, Water, and Infrastructural Decay. *Environment and Security*, Vol. 1. Nos. 3-4. 187-208.

Westing, A. 1972. Herbicides in War: Current Status and Future Doubt. *Biological Conservation*, Vol. 4. No. 5.

Wester, P. Mishra, A. Mukherji, A. and Shrestha, A. 2019. The Hindu Kush Himalaya Assessment: Mountains, Climate Change, Sustainability and People. In *ICIMOD (International Centre for Integrated Mountain Development)*. Cham: Springer.

WFP (World Food Program). 2023. *Drought in the Horn of Africa*. July 2023.

WHO (World Health Organization). 2020. *Vector-borne Diseases*. [online] Available at: https://www.who.int/news-room/fact-sheets/detail/vector-borne-diseases [2023-04-28].

WHO (World Health Organization). 2022. *Drinking-water*. [online] Available at: https://www.who.int/news-room/fact-sheets/detail/drinking-water [2023-03-10].

WHO and UNICEF. 2021. *Progress on Household Drinking Water, Sanitation and Hygiene: 2000-2020. Five Years into the SDGs. WHO/UNICEF Joint Monitoring Program for Water Supply, Sanitation and Hygiene*. Geneva: WHO and UNICEF.

WMO (World Meteorological Organization). 2021a. *2021 State of Climate Services: Water*. WMO-NO.1278.

WMO (World Meteorological Organization). 2021b. *WMO Atlas of Mortality and Economic Losses from Weather, Climate and Water Extremes (1970-2019)*. WMO-No.1267. Geneva: WMO.

WMO (World Meteorological Organization). 2022. *United in Science 2022: A Multi-Organization High-Level Compilation of the Most Recent Science Related to Climate Change, Impacts and Responses.*

WMO (World Meteorological Organization). 2023a. *State of the Global Climate 2022.* WMO-No.1316.

WMO (World Meteorological Organization). 2023b. *Provisional State of the Global Climate in 2023.*

WMO (World Meteorological Organization). 2023c. *State of the Climate in Africa, 2022.* WMO-No. 1330. Geneva: WMO.

WMO (World Meteorological Organization) and CCCS (Copernicus Climate Change Service). 2023. *State of the Climate in Europe 2022.* WMO-No.1320. Geneva: WMO.

Wolf, A.T. 1998. Conflict and Cooperation along International Waterways. *Water Policy*, Vol. 1. No. 2. Pp. 251-265.

Wood, J. 2023. *The Climate Crisis Disrupts 40 Million Children's Education Every Year. Here's How We Could Fix It.* World Economic Forum. 14 February 2023.

Wood, L. Neumann, K. Nicholson, K. Bird. B. Dowling, C. and Sharma, S. 2020. Melting Himalayan Glaciers Threaten Domestic Water Resources in the Mount Everest Region, Nepal. *Frontiers in Earth Science.* Vol. 8. No. 128.

World Bank. 2007. *Lao PDR Environment Monitor*, Washington, DC. World Bank.

World Bank. 2020. *Poverty and Shared Prosperity: Reversals of Fortune.* Washington, DC: The World Bank.

World Bank. 2022a. *China: Country Climate and Development Report.* World Bank Group East Asia Pacific. October, 2022.

World Bank. 2022b. *Water.* [online] Available at: https://www.worldbank.org/en/topic/water/overview [2023-03-09].

World Bank. 2022c. *Poverty.* [online] Available at: https://www.worldbank.org/en/topic/poverty/overview [2023-09-25].

Yamashita, N. and Trinh, T-A. 2022. Long-Term Effects of Vietnam War: Agent Orange and the Health of Vietnamese People after 30 Years. *Asian Economic Journal*, Vol. 36. No. 2. Pp. 180–202.

YPCCC (Yale Program on Climate Change Communication). 2022. *Politics & Global Warming.* April 2022.

YPCCC (Yale Program on Climate Change Communication). 2023. *Global Warming's Six Americas.* [online] Available at: https://climatecommunication.yale.edu/about/projects/global-warmings-six-americas/ [2023-08-21].

Zachariah, M. et al. 2023. *Extreme Humid Heat in South and Southeast Asia in April 2023 Largely Driven by Climate Change, Detrimental to Vulnerable and Disadvantage Communities.* Grantham Institute for Climate Change Faculty of Natural Sciences.

Zacharias, M., Philip, S., Pinto, I., Vahlberg, M., Singh, R., Otto, F., Barnes, C. and Kimutai, J. 2023. *Extreme Heat in North America, Europe and China in July 2023 Made Much More Likely by Climate Change.* Grantham Institute for Climate Change.

Zaveri, E., Russ, J., Khan, A., Damania, R., Borgomeo, E. and Jägerskog, A. 2021. *Ebb and Flow, Volume 1: Water, Migration, and Development.* Washington, DC: World Bank.

Zawahri, N. and Mitchell, S. 2011. Fragmented Governance of International Rivers: Negotiating Bilateral versus Multilateral Treaties. *International Studies Quarterly*. 55. Pp. 835-858.

Zhong, Y., Liu, Q., Sapkota, L., Luo, Y., Wang, H., Liao, H. and Wu, Y. 2021. Rapid Glacier Shrinkage and Glacial Lake Expansion of a China-Nepal Transboundary Catchment in the Central Himalayas, between 1964 and 2020. *MDPI*, Vol. 13. No. 18.

Zhang, K., Cao, C., Chu, H., Zhao, L., Zhao, J. and Lee, X. 2023a. Increased Heat Risk in Wet Climate Induced by Urban Humid Heat. *Nature*, Vol. 617. Pp. 738-742.

Zhang, Y., Zheng, H., Zhang, X., Leung, R., Liu, C., Zheng, C., Guo, Y., Chiew, F., Post, D., Kong, D., Beck, H., Li, C. and Blöschl, G. 2023b. Future Global Streamflow Declines Are Probably More Severe Than Previously Estimated. *Nature Water*, Vol. 1. Pp. 261-271.

Zentelis, R. and Lindenmayer, D. 2015. Bombing for Biodiversity- Enhancing Conservation Values of Military Training Areas. *Conservation Letters: A Journal of the Society for Conservation Biology*, Vol. 8. No. 4. Pp. 299-305.

Index

A
Afghanistan, 52, 55, 115
Agent Orange, 49
Air forces, 60
Air pollution, 52
　aggravated, 53
　detrimental, 52
Alaska, 34, 35
Alps, 75, 76
Amazon, 90
'America first' foreign policy, 30
American Clean Energy and Security Act, 40
Andes, 75, 88, 96
Antarctica, 59, 84
Antarctic Treaty, 148
Anthropogenic climate change, 7, 24
Anti-immigration mobilisation, 116
Anti-submarine warfare techniques, 59
Aramco, 37, 44
Arctic Council, 78, 82
Arctic ice sheets, 78
Arctic Ocean, 78–80
　deep-sea mining, 79
　melting sea ice, 78
　overlapping claims in, 81 (figure)
Arctic region, 9
Argentina, 75
Armed conflicts, 6, 13, 50, 51, 94
　chemical toxicity, 52
　deforestation, 51
　detrimental air pollution, 52
　identity-based, 5
　ongoing, 50
　radioactivity, 52
Australia, 38, 63
Austria, 75
Authoritarianism, 133–138

B
Behrman, S., 119
Biden, J., 10, 25, 26, 34, 35, 55, 83
Biodiversity, 41
Blame game, 28, 44
Borders
　conflicts, 70–74
　island, disputes, 83–87
　national, 109–114
　protection, 3
　shifting territories and, 69–88
Botswana, 73
Bushby, J., 133
Buzan, B., 3

C
Canada, 80
　Northwest Passage, 80
　wildfires in, 8
Carbon capture and storage (CCS) technologies, 37
Carbon dioxide (CO_2), 7, 8, 24, 52
Carbon Disclosure Project, 18
Catastrophic flooding, 132
Chagos Islands, 85
Chemical toxicity, 52
Chile, 62, 75
China, 10, 11, 28, 30, 45
　Arctic, scientific expeditions, 82
　coal production, 31
　downstream hydropower dams, 31
　drought, 91
　Europe's dependency, 58
　Polar Silk Road, 83
　renewable energy sources, 31
Chinese Communist Party, 82
Circular carbon economy, 37
Civil society movements, 43
Civil wars, 2, 5
Climate adaptation, 10
Climate Adaptation Plan, 55
Climate apathy, 23–45
Climate change, 1, 7, 9, 11, 13–15, 69, 108
　anthropogenic, 7, 24
　Arctic, 83
　convergence of, 137
　destabilising effects of, 14
　disinformation, 36
　economic developments and, 13
　effects of, 10
　in elections, 33
　environmental destruction and, 47–53
　evidence of, 10
　expenditure, military, 62–66, 64 (figure)
　far-reaching impacts, 10

food production, 92
glaciers, 74
Global South, 133
immediate threats, 13
international discourse on, 11
irreversible effects of, 54
military, 58–62
national policy and, 19
partisan issue, 33, 34
pre-election poll, 33
profound impacts, 10
recognition of, 12
severity of, 36
threat multiplier, 11
water and, 89–90
Climate crisis, 1–21, 125
  global militaries and, 47–67
  human-induced, 7
  multilateral agreements, 15
  political responses to, 23–25, 32–35
  urgency of, 25–27
Climate election, 33
Climate-fuelled human rights violations, 38
Climate funds, 25
'Climategate' scandal, 37
Climate migration, 107–124
  and conflicts, types of, 114–116
  existing responses, 122–123
  national borders, securitisation of, 109–114, 111 (table)
  political crises, 116–118
  populism, rise of, 116–118
  refugee, 118–122
Climate mitigation, 10, 19, 33
Climate negotiations, 25, 28, 42
Climate President, 34, 35
Climate refugees, 118
Climate risk regions, 133–138
Climate security, 21
  as concept, 9–12
  evolution of, 10, 12
  global issues, 11
  as national security, 12–15
  origin of, 10
Club of Rome's Project on the Predicament of Mankind, 12
Cold Wars, 2, 143
  in Europe, 2
  globalisation and, 3, 10
  radiation levels, 50
Commission on Human Security, 4
Conference of the Parties (COP), 25, 40
Conflict resolution, 73

Continental Ice, Argentina, 75
COP15, 151
COP27, 30, 40, 43
COP28, 25, 30, 147
Copenhagen Climate Conference (2009), 42
Copenhagen Summit, 37
Coral reefs, 113
Corporate lobbying, 40
Costa Rica, 72
Cottrell, L., 56
COVID-19 pandemic, 4, 24, 26, 94
Crawford, N., 49, 56
Cyberattacks, 5
Cyberspace, 5

D
Dam construction, 99
Deforestation, 51, 52, 90, 108
Democracy, 125–142
  dilemma, 35
  policymaking, 39
Democratic Republic of Congo (DRC), 51, 58
Dengue fever, 130
Denmark, 27, 80, 81
Department of Defense, 54, 55
Depleted uranium, 49
Desertification, 60, 108
Disinformation spread, 5
Droughts, 60, 71, 130
  economic impact of, 111
  heat stress and, 111
  Somalia, 110

E
Ecological refugees, 118
Economic crises, 2
Economic welfare, 133
EEZ. See Exclusive Economic Zones (EEZ)
Emission Trading System (ETS), 40
Energy-efficient products, 28
Environmental destruction, 47–53
Environmental hazards, 52
Environmental migration, 108
Environmental refugees, 118
European Commission, 41
European Drought Observatory, 91
European Union, 27, 29, 79, 117, 132
  Emission Trading System (ETS), 40
  green transition, 41
  military emissions, 56, 56 (table)
Exclusive Economic Zones (EEZ), 79, 80, 84–87

Extreme weather events, 94, 130
ExxonMobil, 36

F
False information, 39
Finland, 63, 71, 72
Firefighting, 61
Flooding, 10, 60
  Assam, 110
  catastrophic, 132
  Germany, 129
  Libya, 110
  Pakistan, 110, 111, 131
Food insecurities, 94, 136
Food security, 13, 92, 94
Food supply, 93–95
Forced displacement, 121
Fossil fuels
  burning of, 24
  dependency, 63
  industries, 36, 37
  phasing out of, 26
  production, 8, 26
  reduction, 26
  transition risk, 26
France, 10, 33
  climate footprint, 57
  militaries, 57
  Ministry of the Armed Forces, 57
Freedom from fear, 3–4
Freedom from want, 3, 4
'Fridays for Future' movement, 35

G
Ganges water-sharing, 100
Gayoom, M. A., 112
Gaza, 43, 51
Genocide, 2
Geopolitical tensions, Arctic, 83
Germany, 57, 129
Glacial lake outburst floods (GLOFs), 95–97
Glaciers, 69
  Alps, 75
  boundaries, 74–78
  Himalayas, 76
  melting of, 13, 14, 74, 88
Global Climate Risk Index, 125
Global Compact on Refugees, 122
Global instability, 27
Globalisation, 2–4, 10, 20
Global North, 42, 62, 114, 122, 125, 138–141
  blaming, 139

natural hazards, economic losses from, 128
Global politics, 3
Global power, 2
Global Risk Report, 32
Global South, 42, 62, 125
  climate change, 133
  climate insecurity, 128–133
  poorer economies, 131
Global stability, 10, 21
Global village, 6
Global warming, 7–9, 24, 36, 78, 89, 107, 130
Greenhouse gas emissions, 7, 8, 15, 18, 24, 27, 49, 92
  CO_2 emissions, 25
  environmental destruction and, 50
  fossil fuels, 25
  highest level of, 24
  Nationally Determined Contributions, 25
Greening the military, 53–58
Greenland, 84
Green Party, 33
Greenwashing, 18, 38
Gross domestic products (GDP), 128, 129
Groundwater, 93, 101
  management, 103
  surface soil and, 50
  wars, 103

H
Hamas, 51
Heat stress, 111
Heatwaves, 131
Heung River, 70
High floods, 95–97
Himalayas, 69, 75, 76, 96, 102
Hindu Kush Himalayan glaciers, 76
Hotter temperatures, 14, 60
Human-induced climate change, 7
Human Rights Committee (HRC), 119
Human security, 3, 4, 93–95
Hunger, 4
Hurricane Maria, 129
Hurricane Michael, 129
Hybrid threats, 4
Hydropower, 98
Hydro-project destruction, 89, 95–97

I
Income disparity, 133
India, 26, 30
  Bangladesh and, 72

fossil gas, 38
glaciers, 77
national energy plans, 31
Pakistan and, 76
Indian Armed Forces, 62
Indigenous rights, 80, 81
Industrial Revolution, 21, 29
Intergovernmental Panel on Climate
    Change (IPCC), 10, 17, 24, 84, 107
    misinformation, 36
    sea-level rise, 112
    Sixth Assessment Report (2023), 42
Internally displaced persons (IDPs), 121
International Court of Justice, 73
International Displacement Monitoring
    Centre (IDMC), 110
International Energy Agency (IEA), 26
International Labour Organization (ILO),
    131
International Law Commission, 48
International River Boundaries Database,
    70
International Tribunal of the Law of the
    Sea (ITLOS), 85
Internet of Things (IoT), 5
Ioane Teitiota, 119
IPCC. *See* Intergovernmental Panel on
    Climate Change (IPCC)
Iraq, 50, 55
Islamic State (IS), 50
Island borders, 83–87
Israel, 51
Italy, 75, 91

J
Japan, 63, 85
Jinping, X., 25

K
Kent, A., 119
Kerry, J., 147
Kyoto Protocol, 16, 54

L
Land use changes, 7, 8
Large-scale cross-border migration, 114
Large-scale population migration, 13
Law of the Sea, 148
Libya, 95, 110
Lithuania, 63
Lobbying, 39–42
Long-term Low Emissions and
    Development Strategies (LT-LEDS),
    17, 26

Loss-and-Damage fund, 43
Low- and middle-income countries, 92

M
Malaria, 130
Maldives, 62, 85
Mauritius, 85, 86
Mekong River, 70
Melting snow, 78–83, 81 (figure)
Methane (CH$_4$), 7, 8, 24
Mexico, 72, 86
Middle East, 9, 37, 50
Middle East and North Africa (MENA), 111,
    112
Migration Crisis (2015), 117
Militaries, 47–67
    climate change impacts, 58–62
    emissions, in EU countries, 56, 56 (table)
    environmental destruction and, 47–53
    expenditure and, 62–66, 64 (figure)
    greening, 53–58
    preparedness acts, 50
    previous wars, 48–49
Misinformation, 36–39, 44
Mont Blanc, 75
Multilateral agreements, inadequacy of,
    15–18
Multilateralism, 6
Myanmar, 51

N
Namibia, 73
Napalm, 49
NAS Key West, 59
National borders, securitisation of,
    109–114, 111 (table)
National incentives, 28
Nationally Determined Contributions
    (NDCs), 16, 17, 25, 26
National security, 149–154
Nation-states, 146–148
NATO, 57, 63–65, 82, 88
    energy transition, 58
    Russian invasion of Ukraine, 57
    Scientific Committee, 53
Natural hazards, 110
    Global North, 128
    internal displacements, 111 (table)
Naval Station Norfolk, 59
Nepal, 137
Netherlands Ministry of Defense, 57
Net-zero emissions, 30, 37, 38
New Zealand, 119
Nicaragua, 72, 73

Nigeria, 26
Non-governmental organisations (NGOs), 39, 147
Non-military threats, 2
Non-Russian fossil fuels producers, 27
Non-state actors, 11, 21
North Africa, 9, 50
Northern Sea Route, 79
Northwest Passage, 79, 80
Norway, 27, 81, 129
Norwegian oil, 33
Notre Dame Global Adaptation Initiative (ND-GAIN), 126
Nuclear weapons, adverse effects of, 50

O
Obama, B., 55
Only one Earth, 15
Organization for Economic Co-operation and Development (OECD), 54, 56
Outer Space Treaty, 148

P
Pacific Island Forum, 87
Pacific Ocean, 86, 127
Pakistan, 76, 110, 111, 115, 131
Panama Canal, 79
Paris Agreement, 16, 17, 24–27, 30, 38, 40, 54, 138
  climate financing for Global South, 42–43
  low-carbon economy, 42
Parkinson, S., 56
Patagonia, 18
Permafrost, 9, 59–60
Peru, 136
Poland, 63
Polar Silk Road, 83
Political apathy, 23
Political independence, 2
Political stability, 3, 78–83
Pompeo, M., 79
Population displacement, 8, 18, 21, 107–109
Populism, 134
Portugal, 48, 91
Post-Cold War, 2
Post-war reconstruction, 49
Poverty, 131, 133
Private sectors, 18–20
Protectionism, 27–32, 29 (figure), 44
Public funds, 36–39
Public perception, 41–42
Putin, V., 38

R
Racism, 2
Radioactivity, 52
Rare earth metals, 58
Refugee, 118–122
Refugee Convention (1951), 118, 119
Regime legitimacy, 125–142
Reluctance, 19
Renewable energy sources, 28, 57
Renewable resources, 13
Republic of Kiribati, 119
Rhine River, 103
Rio Grande River, 72
River basin organizations (RBOs), 104
River border, 70, 71
Robinson, M., 150
Russia, 10, 11, 29
  Arctic projects, 82
  climate change denying, 38
  energy strategy (2035), 38
  NATO and, 83
  net-zero emissions, 38
  Vostok Oil Project, 79

S
Saudi Arabia
  Aramco, 37
  greenwashing, 38
Scarcity, 13
SDGs. *See* Sustainable Development Goals (SDGs)
Sea-level rise, 8, 13, 14, 59, 84, 87, 107, 112
  rate of, 113
  Western Pacific, 113
Second World War, 50
Security, 1–7
  connected risks and, 6
  definition of, 3, 6
  global stability and, 10
  policies and, 2
  threats, 4, 6
Security Council, 10–12
Self-preservation, 23–45
  political apathy and, 23, 43–44
Semey Radiological Institute, 50
Siachen Glacier, 77, 88
Singapore, 126
Sixth Assessment Report (2023), 42
Small Island Developing States (SIDS), 128, 129
Social cohesion, 9
Social movements, 18–20
Soil degradation, 52
Somalia, 110

South Asia, 131
Southern Patagonian Ice Field, Chile, 75
Sovereignty, 3
Soviet Union, 2, 73
 dissolution of, 2
 nuclear tests, 50
Spanish election, 33
State sovereignty, 1, 2
Stockholm Conference (1972), 15
Storm autocracies, 136
Storm Hans, 129
Suez Canal, 79
Sullivan, D., 65
Sustainability, 12
Sustainable Development Goals (SDGs), 4, 94, 130
Sweden, 33, 48, 63, 71, 72
Swiss Alps, 75
Switzerland, 75, 126
Syria, 50, 51, 117

T
Taiwan, 65, 85
Task Force of Displacement, 122
Territorial integrity, 1
Territories, 69–88
Thai-Lao Joint Boundary Commission, 71
Threat multiplier, 11, 84, 102, 150
Thunberg, G., 35
Trans-border migrants, 13
Treaty of Fredrikshamn, 71
Trident Juncture, 83
Tropical forests, 51
Trump, D., 25, 30, 55

U
Ukraine, 43, 52, 53, 94
 European militaries, 61–62
 global instability, 27
 Russia's military invasion of, 26
Ullman, R., 3
Unconventional threats, 2
UNDP. *See* United Nations Development Programme (UNDP)
UN Framework Convention on Climate Change (UNFCCC), 16
United Kingdom, 10, 27, 54, 85
United Nations, 38
 Aramco, 38
 Climate Change Conference (COP21), 16
 Conference on Environment and Development (1992), 16
 Conference on the Human Environment (1972), 16
 Department of Economic and Social Affairs, 4
 Environmental Programme, 17
 Security Council, 10, 71, 84
United Nations Convention on the Law of the Sea (UNCLOS), 79, 84, 85, 87
United Nations Development Programme (UNDP), 3
United Nations Environment Programme (UNEP), 89
United Nations Framework Convention on Climate Change (UNFCCC), 122
United States, 2, 10, 29, 45
 Afghanistan and, 52
 Arctic and, 82, 83
 clean energy economy, 40
 Congress election (2022), 33
 cumulative emissions, 30
 Department of Defense, 59
 firefighting, 61
 fossil fuel-producing companies, 36
 Mexico and, 72
 military, fuel usage, 48
 nuclear tests, 50
 oil production, 34
 Strategic Reserve, 26
 University of Notre Dame, 126
 Vietnam War, 49
 weapons, 55, 56
 Willow Project, 34, 35
University of Notre Dame, 126
Urbanisation, 116
Ussuri River, 73

V
Varieties of Democracies (V-Dem), 135
Vector-borne diseases, 61
Vietnam War, 49, 51
Vulnerability, 51

W
Wallensteen, P., 5
Warfares, 5, 50
Warmer temperatures, 61
War profiteering, 34
Water abundance, 90
Water access, 103–105
Water Conference (2023), 90
Water conflict, 89–106
Water crisis, 90
Water cycle, 89, 90
Water development projects, 98–99
Water insecurities, 94, 111
Water scarcity, 13, 90

agriculture, 91–93
economic development, 91–93
Water-sharing agreements, 99–101, 100 (figure)
Water wars, 89, 101–103
Waxman-Markey Bill, 40
Well-being, 2, 20
Well-functioning welfare systems, 133
Westing, A., 49
Wildfires, 8, 33
Willow Project, 34, 35
Windfall tax, 34
World Bank, 91, 95, 110, 111, 131
World Meteorological Organization (WMO), 100, 128
The world's factory, 31
World Social Report (2020), 4

X
Xenophobia, 2

Y
Yale Program on Climate Change Communication, 32
Yamal Liquified Natural Gas project, 82
Yemen, 95

Z
Zika virus, 130
Zimbabwe, 130

www.ingramcontent.com/pod-product-compliance
Lightning Source LLC
Chambersburg PA
CBHW051546020426
42333CB00016B/2127